Letts

GCSE EXAM SECRETS

PHYSICS

GCSE
Exam
Secrets

Physics

Graham Booth

CONTENTS

To revise any of these topics more thoroughly, see *Letts Revise GCSE Physics Study Guide.*

(see inside back cover for how to order)

THIS BOOK AND YOUR GCSE EXAMS

Introduction

This book is designed to help you get better results.

▶ Look at the grade A and C candidates' answers and see if you could have done better.

▶ Try the exam practice questions and then look at the answers.

▶ Make sure you understand why the answers given are correct.

▶ When you feel ready, try the GCSE mock exam papers.

If you perform well on the questions in this book you should do well in the examination. Remember that success in examinations is about hard work, not luck.

What examiners look for

▶ Examiners are obviously looking for the right answer, however, it does not have to match the wording in the examiner's marking scheme exactly.

▶ Your answer will be marked correct if the answer is correct physics even if it is not expressed exactly as it is on the mark scheme. The examiner has to use professional judgement to interpret your answers. You do not get extra marks for writing a lot of irrelevant words.

▶ You should make sure that your answer is clear, easy to read and concise.

▶ You must make sure that your diagrams are neatly drawn. You do not need to use a ruler to draw diagrams. Diagrams are often drawn too small for the examiner to see them clearly. They should be clearly labelled with label lines.

▶ You will often have to draw a graph. At Foundation tier axes of the graph and scales are given. However, at Higher tier you usually have to choose them. Make sure you always use over half the grid given and that you label the axes clearly. If the graph gives points that lie on a straight line use a ruler and a sharp pencil for this line. If it is a curve, draw a curve rather than join the points with a straight line. The points may not all fit on the best curve or straight line. Balance your curve or line so that there are as many points above it as there are below it. Remember you may have some anomalous points and your line or curve should not go through these.

Exam technique

▶ You should spend the first few minutes of the examination time reading through the whole question paper.

▶ Use the mark allocation to guide you on how many points you need to make and how much to write.

▶ You should aim to use one minute for each mark; so if a question has 5 marks, it should take you 5 minutes to answer the question.

▶ Plan your answers; do not write down the first thing that comes into your head. This is absolutely necessary in questions requiring continuous and extended prose.

▶ Do not plan to have time left over at the end. If you do, use it usefully. Check you have answered all the questions, check arithmetic and read longer answers to make sure you have not made silly mistakes or missed things out.

DIFFERENT TYPES OF QUESTIONS

Questions on end of course examinations (called terminal examinations) are generally structured questions. Approximately 25% of the total mark, however, has to be awarded for answers requiring extended and continuous prose.

Structured questions

A structured question consists of an introduction, sometimes with a table or a diagram followed by a number of parts, each of which may be further sub-divided. The introduction provides much of the information to be used, and indicates clearly what the question is about.

▶ Make sure you read and understand the introduction before you tackle the question.

▶ Keep referring back to the introduction for clues to the answers to the questions.

Remember the examiner only includes the information that is required. For example, if you are not using data from a given table in your answer you are probably not answering it correctly.

Structured questions usually start with easy parts and get harder as you go through the question. You do not have to complete each part before you tackle the next part.

Some questions involve calculations. Where you attempt a calculation you should always include your working. Then if you make a mistake the examiner might still be able to give you some credit. It is also important to include units with your answer.

Some structured questions require parts to be answered with longer answers. A question requiring continuous writing needs two sentences with linked ideas. A question requiring extended writing may require an answer of six to ten lines. Candidates taking GCSE examinations generally do less well at questions requiring extended and continuous writing. They often fail to include enough relevant scoring points and often get them in the wrong order. Marks are available on the paper here for Quality of Written Communication (QWC).
To score these you need to write in correct sentences, use scientific terminology correctly and to sequence the points in your answer correctly.

 This logo in a question shows that a mark is awarded for QWC.

Approximately 5% of the marks for the examination as a whole are awarded for questions testing Ideas and Evidence. On modular courses this can form 10% of the marks on the terminal papers, since this is not tested on coursework or module tests. Usually these questions do not require you to recall knowledge. You have to use the information given to you in the question.

WHAT MAKES AN A/A*, B OR C GRADE CANDIDATE

Obviously, you want to get the highest grade that you possibly can. The way to do this is to make sure that you have a good all-round knowledge and understanding of Physics.

GRADE A* ANSWER

The specification identifies what an A, C and F candidate can do in general terms. Examiners have to interpret these criteria when they fix grade boundaries. Boundaries are not a fixed mark every year and there is not a fixed percentage who achieve a grade each year. Boundaries are fixed by looking at candidates' work and comparing the standards with candidates of previous years. If the paper is harder than usual the boundary mark will go down. The A* boundary has no criteria but is fixed initially as the same mark above the A boundary as the B is below it.

GRADE A ANSWER

A grade candidates have a wide knowledge of physics and can apply that knowledge to novel situations. An A grade candidate generally gains a high proportion of the marks available on each question. A grade A can only be achieved on the Higher tier papers. The minimum percentage for a grade A is about 70%.

GRADE B ANSWER

B grade candidates have a reasonable knowledge of the topics identified as Higher tier only. The minimum percentage for a B candidate is exactly halfway between the minimum for A and C (on Higher tier).

GRADE C ANSWER

C grade candidates can get their grade either by taking Higher tier papers or by taking Foundation tier papers. There are some questions common to both papers and these are aimed at C and D candidates. The minimum percentage for a C on Foundation tier is approximately 60% but on Higher tier it is approximately 40%.

If you are likely to get a grade C or D on Higher tier you would be seriously advised to take Foundation tier papers. You will find it easier to get your C on Foundation tier as you will not have to answer the questions targeted at A and B.

HOW TO BOOST YOUR GRADE

Grade booster ⋯⋙ How to turn C into B

▶ All marks have the same value. Don't forget the easy marks are just as important as the hard ones. Learn the relationships between physical quantities – these are easy marks in exams and they reward effort and good preparation. If you want to boost your grade, you **cannot** afford to miss out on these marks – they are easier to get.

▶ Look carefully at the command word at the start of the sentence. Make sure you understand what is required when the word is **state, suggest, describe, explain** etc.

▶ In numerical calculations, always include units with the value of each physical quantity.

▶ Use the names of physical quantities rather than symbols, this way you are less likely to make an error.

▶ Read the question twice and underline or highlight key words in the questions. E.g. Calculate the _power_ of an appliance if the _current_ passing in it is _5.5 A_ when the voltage across it is _240 V_.

▶ Make sure that you use any data that is given in the question.

Grade booster ⋯⋙ How to turn B into A/A*

▶ Make sure that you know all the relationships between physical quantities and all the units. Practise using these relationships.

▶ Make sure that you always give the correct unit with any physical quantity. Do not leave this to chance. Make a list of physical quantities such as acceleration and current. Test yourself by working through the list, writing down the correct unit for each quantity.

▶ Check any calculations you have made at least twice, and make sure that your answer is sensible. For example, if you divide 0.49 by 1.9 the answer is approximately 0.25.

▶ In calculations, give your answer to the same number of significant figures as the given data.

▶ Give full reasons in questions that require you to "explain". If you are hesitant about something, put it down anyway. You do not lose marks for wrong information, unless it contradicts another part of your answer.

▶ In questions requiring extended writing make sure you make enough good points and you don't miss out important points. Read the answer through and correct any spelling, punctuation and grammar mistakes.

▶ For A/A* you have to show that you have a really good understanding of physics and are fluent with physics terms. You need a good grip of topics such as the diffraction of waves and electromagnetic induction.

Electricity

To revise this topic more thoroughly, see Chapter 1 in *Letts Revise GCSE Physics Study Guide.*

 Try this sample GCSE question and then compare your answers with the Grade C and Grade A model answers on page 10.

The diagram shows how an electrical convector heater is connected to the mains supply.

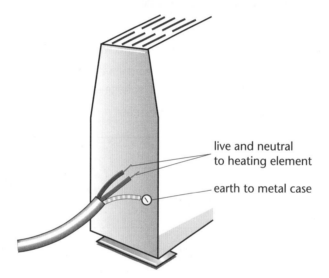

live and neutral to heating element

earth to metal case

When operating from the 240 V mains supply, the current in the heating element is 8.4 A.

a Calculate the resistance of the heating element.

....

....

.... **[3]**

b Calculate the power of the heating element.

....

....

.... **[3]**

c Which of the wires connecting the heater to the mains supply:

   (i) Supplies the energy input to the heater?

.... [1]

   (ii) Acts, together with the fuse, as a safety device to protect against electrocution?

.... [1]

d Energy from the mains electricity supply is charged at the rate of 7p for each kilowatt-hour (kWh).

   (i) Calculate the cost of using the heater between 8 am and 2 pm.

....

.... [2]

   (ii) Suggest why the real cost of leaving the heater switched on between 8 am and 2 pm is likely to be less than the answer to **d (i)**.

....

.... [2]

(Total 12 marks)

These two answers are at Grade C and A. Compare which one your answer is closest to and think how you could have improved it.

GRADE C ANSWER

Shamba is awarded two marks out of three for correct recall of the relationship and substituting the values. Unfortunately she does not know the unit of resistance, so she loses the third mark.

Although this answer is partially correct (Shamba has just neglected to multiply by the power of the heater) it gains no marks. Had she written down the relationship correctly, she would have had a better chance of gaining marks with a correct substitution and final answer. But she should have realised that it cannot cost £42 to operate a convector heater for 6 hours.

Shamba

a 240 ÷ 8.4 = 28.6 W ✓✓

b 240 × 8.4 = 2016 J ✓✓

c (i) This is the live wire. ✓

(ii) The neutral wire is neutral, so it does not have a charge. ✗

d (i) 8 am to 2 pm is 6 hours and 6 × 7 = 42, so it must be £42. ✗

(ii) Somebody could have turned it off, or electricity is cheaper than 7p. ✗

Similarly to a, Shamba's answer is numerically correct but the unit is given as J instead of W, so again, two marks out of three.

Correct – one mark.

This response reveals confusion between the mains electricity supply and electrostatic charge. No marks are awarded.

No marks. The correct answer is that the heater is fitted with a thermostat that turns it off when the desired temperature is reached, so the heater was not operating for the full 6 hours.

5 marks = Grade C answer

Grade booster ····▷ move a C to a B
Grade C candidates are often uncertain about the formulae for relationships between physical quantities and the correct units for quantities such as power and resistance. Make sure that you learn all the relationships that are required knowledge and you know the correct units for physical quantities.

GRADE A ANSWER

Full marks for a correct calculation and answer. Notice how Mo always gives the correct unit with physical quantities. This is not essential (except in the final answer) but it is good practice.

Wrong. Energy flows along the live wire.

Mo hasn't a clue and is just guessing. Although she does not gain any marks, she was right to attempt an answer, as she cannot lose anything by doing this.

Mo

a Resistance = voltage ÷ current ✓
= 240 V ÷ 8.4 A ✓ = 28.6 Ω ✓

b Power = current × voltage ✓
= 8.4 A × 240 V ✓
= 2016 W ✓

c (i) Earth ✗

(ii) Live ✗

d (i) 6 h × 2.016 kW × 7p/kWh ✓
= 84.7p ✓

(ii) Because during the day the heater uses less electricity than at night. ✗

Full marks again. Grade A candidates should be able to recall relationships and use them to calculate the size of physical quantities.

Wrong again. The earth wire acts with the fuse to prevent electrocution.

Full marks here.

8 marks = Grade A answer

Grade booster ····▷ move A to A*
Your knowledge has to be absolutely rock solid. See how Mo has no difficulty with the calculations, but she falls down on knowing about the functions of the wires in an electricity mains lead. Her other error is in the last part of the question. Common sense should tell her that it cannot have been operating for the complete time interval.

1 A student compares three different metal wires to see which is the best conductor of electricity. He measures the voltage needed to pass a current of 0.8 A in each wire in turn.
The table shows his results.

Wire	Voltage in V
A	5.2
B	0.6
C	12.4

a) Which wire is the best conductor of electricity? Explain your choice.

....

.... ②

b) Calculate the resistance of wire B.

....

....

.... ③

c) While doing the experiment, the student notices that one of the wires gets hot. Calculate the power in each wire and use your answers to explain which wire gets hot.

....

....

....

....

.... ⑤

d) Calculate the quantity of electric charge that passes through each wire in one minute.

....

....

.... ③

TOTAL 13

2 The following circuit is used to investigate how the current in a resistor changes when the voltage across it is increased.

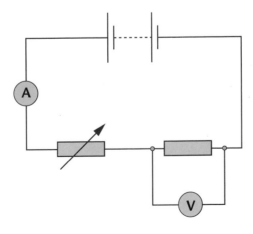

The table shows the results.

Voltage in V	Current in A
0.00	0.00
0.45	0.35
0.95	0.70
1.60	1.20
2.05	1.55
2.40	1.80

a) Use the grid to plot a graph of current against voltage. ④

b) i) Write down the value of the current in the resistor when the voltage across it is 1.20 V.

   .... ①

   ii) Calculate the resistance of the resistor when the voltage across it is 1.20 V.

   ....
   ....
   .... ③

c) A second identical resistor is added to the circuit in series with the existing one.
   Add a line to the graph to show how the current in the circuit varies as the voltage is
   changed. ②

TOTAL 10

3 The graph shows how the current in a filament lamp changes as the voltage is increased.

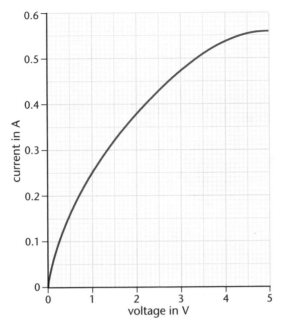

a) Calculate the resistance of the filament when the voltage is 1.5 V.

....

....

.... ③

b) Describe how the resistance of the filament changes as the voltage is increased. Justify your answer by referring to data from the graph.

....

.... ②

TOTAL 5

4

a) Three copper wires are used to connect a metal-cased kettle to the mains supply. They are called live, neutral and earth.

   i) When the kettle is switched on and working normally, current passes in two of these wires. Which two?

   .... ①

   ii) Which wire is connected to the metal case?

   .... ①

   iii) Which wire is connected to the fuse in the plug?

   .... ①

   iv) Each wire is covered in a layer of PVC.
   All three wires are enclosed in an outer PVC layer.
   Explain the purpose of the PVC.

   ....

   .... ②

b) The current passing in the kettle element is 9.5 A. The mains voltage is 240 V. Calculate the power of the heating element.

....

....

.... ③

TOTAL 8

5 When an aircraft is being refuelled, the fuel can become negatively charged as it flows along the metal pipe to the fuel tanks.

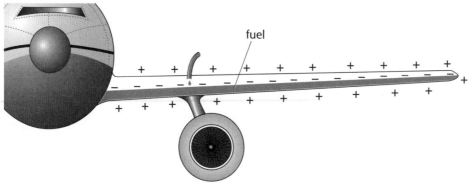

fuel

a) Explain how the fuel becomes negatively charged.

....

.... ②

b) What type of charge does the metal pipe gain?

.... ①

c) Explain why a build-up of charge on the wings of the aircraft could be dangerous.

....

....

.... ④

d) During refuelling, the metal airframe is connected to the ground. Explain how this makes the process safer.

....

.... ②

TOTAL 9

---

6 Electricity retailers sell energy to domestic customers in units of kWh.
The cost of each kWh is around 7p.
In a household, an 8.4 kW shower is in use for 0.6 hours each day.

a) Calculate the energy transfer and the cost of that energy in a 90-day period.

....

....

.... ③

b) Calculate the current in the heating element when it is operating from the 240 V mains supply.

....

....

.... ③

c) The electric circuit to the shower is fitted with a circuit breaker rated at 40 A.

  i) What danger does the circuit breaker **on its own** guard against?

  .... ①

  ii) State two advantages of using a circuit breaker rather than a fuse.

  ....

  .... ②

d) There are three conductors in the cable to the shower. These are called earth, live and neutral.

   i) Describe the function of the live and neutral conductors.

....

.... ②

   ii) Explain how the earth wire, along with the circuit breaker, protects the user from electrocution.

....

.... ③

TOTAL 13

7

a) A variable resistor can be used to change the amount of current passing in a circuit.
The diagram shows a circuit that can be used to illuminate the instruments on a car dashboard.

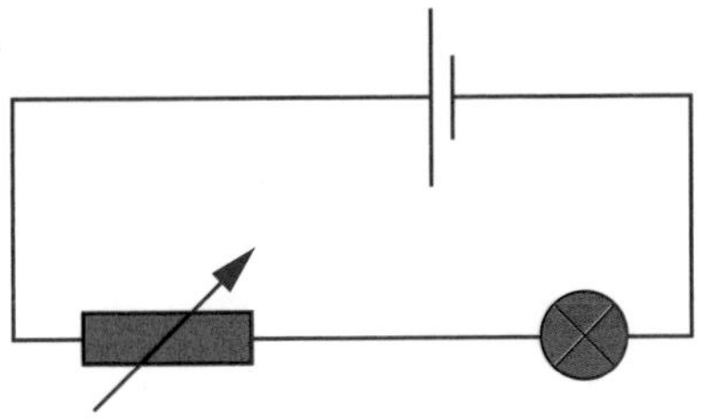

   i) Describe what the circuit is able to do.

.... ①

   ii) Use words from the following list to complete the sentences.

current   **greatest**   **resistance**   **smallest**

When the lamp is dim the........................... of the variable resistor is at its ....

This causes the .... in the lamp filament to be at its........................... .   ④

b) When the lamp is operating at its normal brightness the voltage across the filament is 12 V and the current passing is 0.2 A.

   i) Calculate the resistance of the filament at its normal brightness.

....

....

.... ③

   ii) Calculate the power of the lamp at its normal brightness.

....

....

.... ③

TOTAL 11

8 The diagram represents the electrolysis of a salt solution.

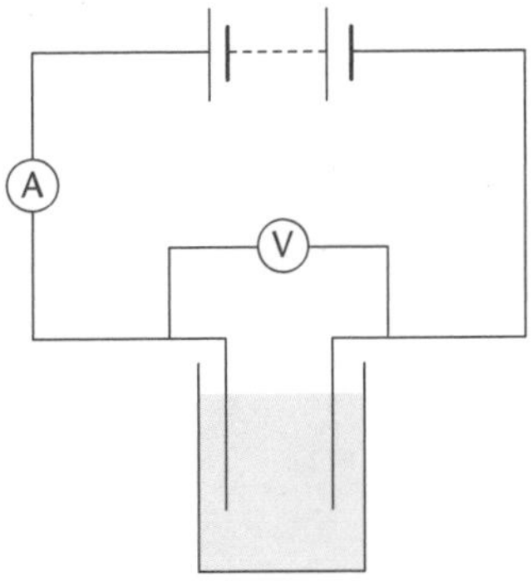

During electrolysis current passes in both the copper wires and in the salt solution.

a) i) Describe the movement of charged particles in the wire.

....

....

.... ③

ii) Describe the movement of charged particles in the solution.

....

.... ②

b) The current in the solution is 0.65 A.
Calculate the quantity of charge that flows through the solution in 120 s.

....

....

.... ③

c) The voltmeter reading is 9.0 V.
How much energy is transferred to the solution in 120 s?

....

....

.... ③

TOTAL 11

9

a) Use the three sets of axes to sketch how current varies with voltage for:

i) a fixed value resistor

ii) a filament lamp

iii) a diode.

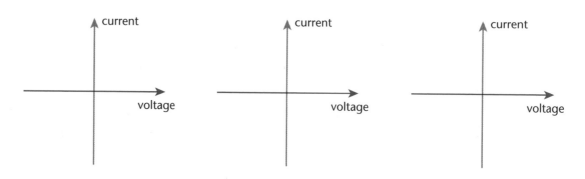

i) a fixed value resistor   ii) a filament lamp   iii) a diode ③

b) The table gives data about the resistance of a thermistor at different temperatures.

Temperature in °C	Resistance in Ω
20.0	64
31.5	45
48.0	27
57.5	21
72.5	13

i) Use the grid to draw a graph of resistance against temperature for the thermistor.

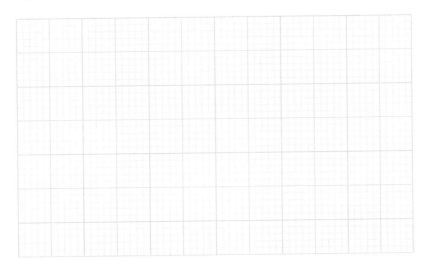

④

ii) Describe how the resistance of the thermistor changes when its temperature changes.

....

.... ②

The thermistor is used in an incubator to maintain the temperature at a constant 37°C.

iii) What is the resistance of the thermistor at 37°C?

.... ①

The thermistor is used in the following circuit.

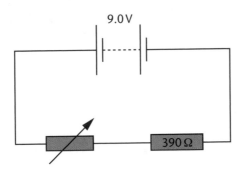

9.0 V

390 Ω

iv) Calculate the current in the circuit when the temperature of the thermistor is 37°C.

....

....

.... ③

v) Calculate the voltage across the thermistor when its temperature is 37°C.

....

....

.... ③

vi) Explain how the voltage across the thermistor changes when its temperature rises.

....

.... ②

TOTAL 18

1 a) B is the best conductor ❶
Because it needs the smallest voltage to
cause the same current. ❶

b) $R = V \div I$ ❶
   $= 0.6\ V \div 0.8\ A$ ❶
   $= 0.75\ \Omega$ ❶

Examiner's tip

You must have the correct unit for the third mark. If your answer is 0.75 with no unit or the wrong unit, you gain two marks out of the three. Although you would usually gain full marks for the correct answer with no working, showing your working allows you to gain some credit, e.g. for knowledge of the formula, even if your final answer is wrong.

c) $P = I \times V$ ❶
The power in wire A, $P = 0.8\ A \times 5.2\ V$ ❶
$= 4.16\ W$ ❶
Similarly for wire B, $P = 0.48\ W$, and for
C, $P = 9.92\ W$ ❶
C gets hottest because it has the greatest
power ❶

Examiner's tip

When you are asked to 'explain why', you must give a reason. The reason in this case is 'because it has the greatest power'.

d) $Q = I \times t$ ❶
For each wire $Q = 0.8\ A \times 60\ s$ ❶
$= 48\ C$ ❶

Examiner's tip

As the wires carry the same current, the same quantity of charge flows through each in one minute. For the formula to be valid, the current must be in amps and the time in seconds.

2 a) Here is the completed graph:

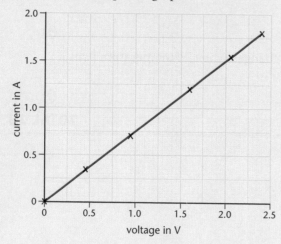

Marks are awarded for:
Correct choice of scales and labelling
of axes ❶
Plotting of points (allow 1 mark for up to
two errors) ❷
Drawing best-fit straight line ❶

b) i) 0.90 A ❶
   ii) resistance = voltage ÷ current ❶
       $= 1.20\ V \div 0.90\ A$ ❶
       $= 1.33\ \Omega$ ❶

Examiner's tip

1.33 is a rounded value. Always give answers to the same number of significant figures as the data.

c) Your line should be less steep than the one
you drew in a). ❶
It should pass through the origin and the
point (2.40, 0.90). ❶

Examiner's tip

There is one mark here for appreciating that increasing the resistance in the circuit reduces the current. The second mark is for realising that the resistance is doubled, so the current is halved.

3 a) resistance = voltage ÷ current ❶
       $= 1.5\ V \div 0.32\ A$ ❶
       $= 4.7\ \Omega.$ ❶

b) The resistance increases ❶
A second calculation to show this ❶

Examiner's tip

Always make it clear how you have taken readings from a graph. In this case you need to read off the current at a voltage that is greater than 1.5 V. Choose a suitable voltage and draw a line up from this value to the curve, then across to the current axis to show how you have found the value of the current.

4 a) i) Live and neutral ❶
     ii) Earth ❶
     iii) Live ❶
     iv) PVC is an insulator ❶
         It prevents the user from coming into
         contact with the conductors/getting
         a shock ❶

Examiner's tip

The question asks you to explain the purpose of the PVC. This means that you must state its purpose and give the reason.

b) power = current × voltage   **1**
    = 9.5 A × 240 V   **1**
    = 2280 W   **1**

5 a) The fuel gains electrons   **1**
    From the metal pipe   **1**
b) Positive   **1**

Examiner's tip

All electrostatic charging is due to the transfer of electrons between objects. The outermost electrons in atoms and molecules are easily removed by friction forces when objects rub together. A common error by GCSE candidates is to state that objects become positively charged by gaining protons.

c) The wings become charged to a high voltage   **1**
    This ionises the air between the wings and the ground   **1**
    The spark caused by charge passing through the air ignites the fuel   **1**
        Quality of written communication mark – the answer is relevant to the question   **1**
d) Electrons travel from the ground, along the wire   **1**
    To neutralise the positive charge   **1**

Examiner's tip

Earthing allows the movement of electrons in either direction along the earth connection. It is equally effective at preventing the build-up of both positive and negative charge.

6 a) energy transfer = power × time
        = 8.4 kW × 0.6 h × 90   **1**
        = 453.6 kWh   **1**
    cost = 453.6 kWh × 7p/kWh
        = 3175.2p or £31.752   **1**

Examiner's tip

The energy formula is not one that you are required to know. It will be given in the question or on a separate formula sheet. The same formula is used for calculating energy transfer in joules when the power is in watts and the time in seconds.

b) current = power ÷ voltage   **1**
    = 8400 W ÷ 240 V   **1**
    = 35 A   **1**
c) i) Overheating of the supply cables   **1**
  ii) A circuit breaker acts to break the circuit in a shorter time than a fuse if a fault occurs   **1**
     A circuit breaker is easily reset   **1**
d) i) The live conductor carries energy to the shower   **1**
    The neutral completes the circuit   **1**

ii) The earth wire is connected to any exposed metal parts of the shower   **1**
    If these become live there is a low-resistance path to earth   **1**
    The resulting high current causes the circuit breaker to break the circuit   **1**

Examiner's tip

Always try to use the correct technical and scientific terms; in this case it is correct to state that the circuit breaker 'breaks the circuit' rather than 'cuts off the power'. What it actually cuts off is the voltage supply.

7 a) i) The circuit allows the brightness of the lamp to be changed   **1**

Examiner's tip

Answers such as 'make the lamp dimmer' or 'make the lamp brighter' would be acceptable.

  ii) Resistance   **1**
     Greatest   **1**
     Current   **1**
     Smallest   **1**

Examiner's tip

Resistance measures the opposition to current; the more resistance in a circuit, the less current passes.

b) i) resistance = voltage ÷ current   **1**
        = 12 V ÷ 0.2 A   **1**
        = 60 Ω   **1**
  ii) power = current × voltage   **1**
        = 0.2 A × 12 V   **1**
        = 2.4 W   **1**

8 a) i) In the metal negatively-charged   **1**
     Electrons   **1**
     Move from negative to positive   **1**
  ii) In the salt solution the charge flow is due to positively-charged ions and negatively-charged ions   **1**
     Moving in opposite directions   **1**

Examiner's tip

In metals only the free electrons can move but in molten or dissolved electrolytes and in ionised gases both the positive and the negative ions move to form a current.

b) $Q = I \times t$ **1**
    = 0.65 A × 120 s   **1**
    = 78 C   **1**
c) energy transfer = voltage × charge
    or $E = V \times Q$ **1**
    = 9.0 V × 78 C   **1**
    = 702 J   **1**

Electricity

Examiner's tip

> The voltage across a power supply or a component is the energy transfer for each coulomb of charge that flows through it. In the case of a power supply the energy is transferred to the charge but as it flows through a component the energy is transferred from the charge.

9 a) The sketch graphs should look like this:

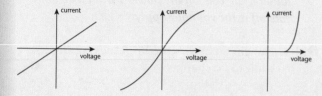

1 mark for each correct graph **3**

b) i) Here is the completed graph:

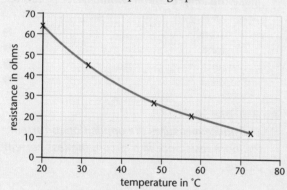

Marks are awarded for:
Correct choice of scales and labelling of axes **1**
Plotting of points (allow 1 mark for up to two errors) **2**
Drawing appropriate curve **1**

Examiner's tip

> When drawing a best-fit line or curve, try to balance the points above the curve with those below the curve. If the graph is a straight line, do not just join the first and last points – these are no more reliable than any other points.

ii) The resistance decreases **1**
As the temperature increases **1**

Examiner's tip

> When describing the relationship between two variables, you need to make clear whether the value of each variable is increasing or decreasing.

iii) 38 Ω **1**
iv) Current = voltage ÷ resistance **1**
= 9.0 V ÷ 428 Ω **1**
= 2.1×10^{-2} A **1**
v) Voltage = current × resistance **1**
= 2.1×10^{-2} A × 38 Ω **1**
= 0.80 V **1**
vi) The resistance of the thermistor decreases but that of the fixed resistor stays the same. **1**
The thermistor has a smaller share of the voltage, so the voltage across it falls. **1**

Examiner's tip

> When two resistors are connected in series, the voltage is shared in the ratio of the resistances.

CHAPTER 2

Force and motion

To revise this topic more thoroughly, see Chapter 2 in *Letts Revise GCSE Physics Study Guide*.

 Try this sample GCSE question and then compare your answers with the Grade C and Grade A model answers on pages 22 and 23.

A car driver notices a hazard in the road ahead. The graph shows how the speed of the car changes from when she sees the hazard.

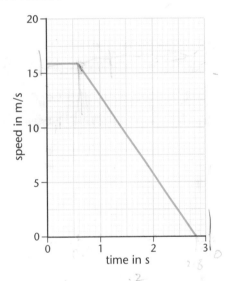

a (i) How long does it take for the driver to react to the hazard?

.... **[1]**

(ii) If the driver's 'reaction time' is doubled, how does this affect the distance that the car travels during this time?

Give an explanation for your answer.

....

.... **[2]**

b Use the graph to calculate the stopping distance of the car.

....

....

.... **[3]**

c (i) Calculate the deceleration (negative acceleration) of the car when it is braking.

....

....

.... **[3]**

(ii) The driver has a mass of 75 kg.

Calculate the size of the force needed to cause the driver to have the same deceleration as the car.

....

....

.... **[3]**

(iii) After the car has stopped, the driver continues to move forwards as the seat belt stretches.

Explain how this stretching of the seat belt reduces the size of the force exerted on the driver.

....

....

.... **[3]**

(Total 15 marks)

These two answers are at Grade C and A. Compare which one your answer is closest to and think how you could have improved it.

GRADE C ANSWER

One mark for a correct reading from the graph.

George W has read the initial speed from the graph, but not realised that he needs to do separate calculations for when the car travels at a constant speed and when it decelerates. No marks.

George W is now showing confusion. Like many candidates at this level, he has no real understanding of $F = m \times a$, so instead he calculates the weight of the driver. No marks.

Given George W's level of understanding, it is not surprising that he cannot explain this in terms of the reduced deceleration and hence less force.

George W

a (i) 0.6 s. ✓

  (ii) It will go twice as far ✓

b distance = speed × time

       = 16 × 2.75 = 44 m. ✗

c (i) change in speed ÷ time ✓

       = 16 ÷ 2.75 = 5.8 m/s

  (ii) force = mass × g

       = 75 × 10 = 750 kg ✗

  (iii) If the seat belt stretches all the force is pushing the belt forwards and not backwards on the driver. ✗

3 marks = Grade C answer

One mark out of two here. George W has given a correct answer but no explanation, which the question specifically asks for.

George W has the right idea. He has made two errors. One is in not working out the time that the car was decelerating. The second is the wrong unit of acceleration – a common error at GCSE level. He gains one mark out of three. Had he not shown his working, he would have gained no marks for a wrong answer and unit.

Grade booster ····➤ move a C to a B

Grade C candidates can usually respond to a straightforward question based on $F = m \times a$, but their understanding of this relationship is very limited. This relationship needs to be understood, and all units given correctly.

Tony

a (i) It takes the driver 0.6 s to react. ✓

Tony has read the graph correctly – one mark.

(ii) The car will travel twice as far, because at a constant speed, the distance travelled is proportional to the time. ✓✓

An excellent answer with a fully correct explanation – two marks.

b This is represented by the area between the graph line and the time axis.

$16 \text{ m/s} \times 0.6 \text{ s} = 9.6 \text{ m}$ ✓

$\frac{1}{2} \times 16 \text{ m/s} \times 2.15 \text{ s} = 17.2 \text{ m}$ ✓

Adding these together gives a distance of 26.8 m. ✓

Tony gains full marks again. Notice how he has made two separate calculations here. One for the distance travelled at constant speed and one for the distance travelled while the car was decelerating. By adding these together, he achieves the correct final answer.

c (i) deceleration = decrease in speed ÷ time ✓

$= 16 \text{ m/s} \div 2.75 \text{ s} = 5.8 \text{ m/s}^2$ ✓

Only two marks out of three here. Tony has made an error in using the wrong time. The actual time of the deceleration is 2.15 s. Fortunately he has shown all his working, so he gains credit for the things that he has done correctly.

(ii) $F = m \times a$ ✓

$= 75 \text{ kg} \times 5.8 \text{ m/s}^2$ ✓ $= 435 \text{ N}$ ✓

Although Tony's answer to **c (i)** is wrong, he has made correct use of this wrong answer to work out the force. He gains full marks for this part of the question. The correct answer is 558 N.

(iii) It takes longer for her to stop, so less force is needed. ✓

Tony gains one mark out of three. He has made a correct statement that it takes longer for her to stop, but has not explained that this reduces the deceleration and hence the force required.

12 marks = Grade A answer

Grade booster ·····⟩ move A to A*

An A* candidate is expected to be able to apply his/her knowledge and understanding of physics to new situations. In this question, a candidate who gives a correct answer to **c (iii)** is showing the aspects expected of an A* candidate.

Force and motion

1 A sprinter runs in a 100 m race.
The graph represents the motion of the sprinter.

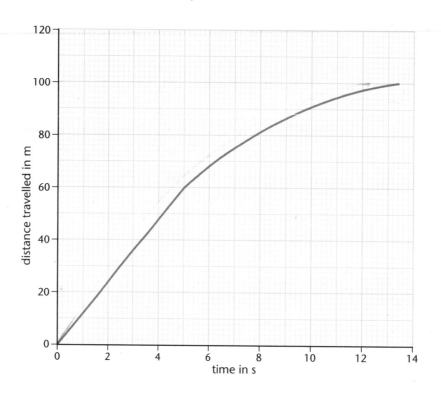

a) How far did the sprinter travel during the first 5 s?

.... ①

b) Calculate the speed of the sprinter during the first 5 s of the race.

....

....

.... ③

c) i) Describe how the speed of the sprinter changed during the last 3 s of the race.

.... ①

ii) Explain how you can tell this from the graph.

.... ①

d) The sprinter who won the race completed the distance in 12.2 s.
Add a line to the graph to represent the motion of the winner of the race. ②

TOTAL 8

2 The graph shows how the distance travelled by a cyclist changes during a cycle ride.

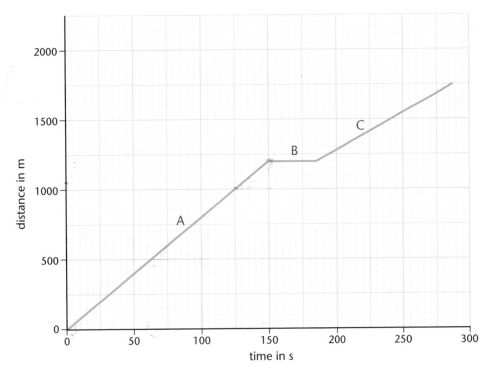

a) On section A of the graph the cyclist travelled at a constant speed. Explain how the graph shows this.

....

.... ②

b) i) Calculate the average speed of the cyclist on the cycle ride.

....

....

.... ③

ii) Explain why this is an average speed.

.... ①

c) When the cyclist brakes, rubber blocks press onto the rims of the wheels.

i) What type of force acts between the rubber blocks and the wheel rims?

.... ①

ii) Explain why the braking of the cyclist is reduced in wet weather.

....

.... ②

iii) What extra precaution should the cyclist take when cycling in wet weather?
Explain why the cyclist should take this precaution.

....

.... ②

TOTAL 11

Force and motion

3 The graph shows how the speed of a parachutist changes after jumping from an aircraft.

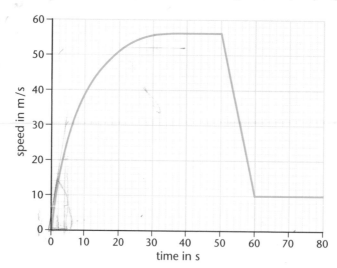

a) During the first 34 seconds shown on the graph, the velocity of the parachutist is increasing. Describe what is happening to the acceleration of the parachutist and explain why this is happening.

....
....
.... ③

b) After 50 seconds, the parachute opens. Calculate the acceleration of the parachutist during the next 10 seconds.

....
....
.... ③

c) i) Explain why the parachutist travels at a constant speed during the period 60 to 80 seconds shown on the graph.

....
....
.... ③

   ii) What name is given to this constant speed?

.... ①

TOTAL 10

4 The diagram shows how the stopping distance of a car is made up of thinking distance and braking distance.

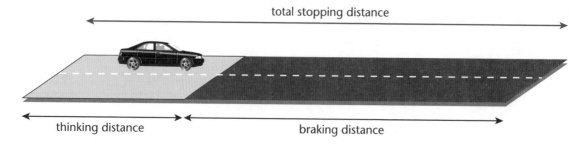

a) i) What is meant by 'thinking distance'?

....
.... ②

ii) State **two** factors that can affect thinking distance.

....

.... ②

(b) i) What is meant by 'braking distance'?

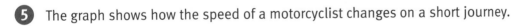

....

.... ③

ii) State **two** factors that can affect braking distance.

....

.... ②

c) A car brakes from a speed of 15 m/s. The graph shows how the speed of the car changes during braking.

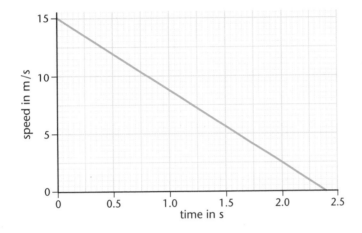

i) Calculate the deceleration (negative acceleration) of the car.

....

.... ③

ii) The total mass of the car and its contents is 850 kg. Calculate the size of the force needed to cause the deceleration.

....

.... ③

TOTAL 15

5 The graph shows how the speed of a motorcyclist changes on a short journey.

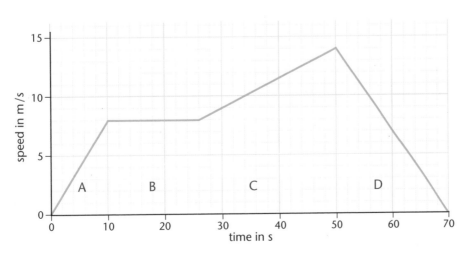

a) i) Describe the motion of the motorcyclist during each stage of the journey.

....

....

....

.... ④

ii) Calculate the acceleration of the motorcyclist during the first 10 s.

....

....

.... ③

iii) The total mass of the motorcycle and motorcyclist is 220 kg. Calculate the force needed to cause the acceleration in the first 10 s shown on the graph.

....

....

.... ③

iv) During the last 20 s shown on the graph the motorcyclist is braking to a halt. Calculate the braking force required.

....

....

....

.... ④

v) The motorcyclist next does a return journey, carrying a passenger who has a mass of 80 kg. If the braking force remains the same, explain how the braking of the motorcycle is affected.

....

.... ②

vi) Suggest **two** other factors that could affect the braking of the motorcycle.

....

.... ②

TOTAL 18

6 The diagram shows a paper clip being held by a magnet.

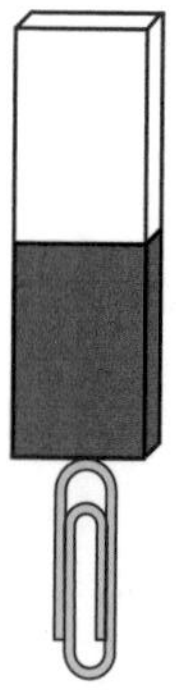

a) Draw two arrows to show the vertical forces on the paperclip and write a short description of each force.

....

.... ④

b) What can you tell about the relative size of the two forces? Give the reason for your answer.

....

.... ②

c) The paperclip is held between two identical magnets in the position shown in the diagram.

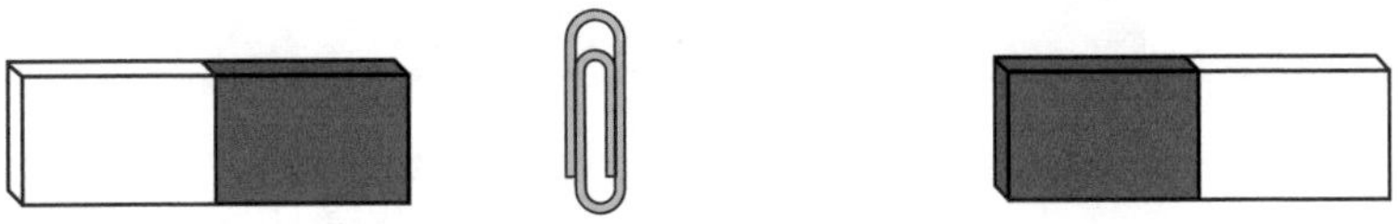

Describe and explain what will happen to the paperclip when it is released.

....

....

.... ③

TOTAL 9

Force and motion

❶ a) 60 m **❶**
b) speed = distance ÷ time **❶**
= 60 m ÷ 5 s **❶**
= 12 m/s **❶**

Examiner's tip

When answering questions that involve calculating physical quantities, always set out your answer as shown above. This way, you gain two out of three marks even if you get the final answer wrong, provided that your method of working is correct. Simply writing down a wrong answer gains no marks at all. Note that to gain the last mark you need both the correct answer and the unit.

c) i) It decreased or became slower **❶**
ii) The gradient or slope decreases **❶**

Examiner's tip

The gradient or slope of a distance–time graph represents the speed of the object. Similarly, the gradient of a velocity–time graph represents acceleration.

d) The line should start at the origin **❶**
And pass through (12.2, 100) **❶**

❷ a) The gradient or slope of the graph represents the speed **❶**
Section A has a constant gradient **❶**
b) i) average speed
= distance travelled ÷ time taken **❶**
= 1750 m ÷ 290 s **❶**
= 6.0 m/s **❶**

Examiner's tip

This is equivalent to working out the slope or gradient of the graph.

ii) The cyclist spent some time not moving; at other times the cyclist moved with different speeds. **❶**
c) i) Friction **❶**
ii) Water gets between the rubber block and the wheel rim **❶**
This reduces the friction force **❶**

Examiner's tip

When a question asks you to 'explain', it is important that you give the full reasons. In this case there are two marks for the question, so you are expected to give two points to gain full marks.

iii) Leave a greater distance between the cycle and any road user in front **❶**
The braking distance is increased **❶**

❸ a) The acceleration is decreasing **❶**

Examiner's tip

The gradient of a speed–time graph represents the acceleration. In this case decreasing, showing that the acceleration is also decreasing.

As the parachutist speeds up the resistive force increases **❶**
The resultant force on the parachutist decreases **❶**

Examiner's tip

The resultant force is the sum of all the forces acting. In this case the parachutist's weight acts downwards and the resistive force acts upwards, so the resultant downward force is equal to (weight – resistive force).

b) acceleration = increase in velocity ÷ time taken **❶**
= (10 m/s – 56 m/s) ÷ 10 s **❶**
= −4.6 m/s² **❶**

Examiner's tip

The acceleration here is negative because the velocity of the parachutist decreased; this negative acceleration is also known as a deceleration.

c) i) The upward push of the air **❶**
Is equal to the downward pull of the Earth **❶**
The forces on the parachutist are balanced or there is no unbalanced force **❶**
ii) Terminal velocity **❶**

Examiner's tip

The difference between speed and velocity is that speed only describes how fast an object is moving. Velocity also describes the direction of motion.

4 a) i) The distance that a vehicle travels ❶
　　　During the driver's reaction time ❶
　　ii) The driver's alertness
　　　Tiredness
　　　The effect of drugs or alcohol
　　　　　　　　any 2, 1 mark each ❷
　b) i) The distance that a vehicle travels from
　　　when the brakes are applied ❶
　　　To when it stops ❶
　　　　Quality of written communication mark,
　　　awarded for accurate spelling (e.g. braking
　　　　rather than breaking), punctuation and
　　　　　　　　　　grammar. ❶
　　ii) Condition of the road surface
　　　Condition of the tyres
　　　Condition of the brakes
　　　Whether the vehicle skids
　　　How hard the brakes are applied
　　　　　　　　any 2, 1 mark each ❷
　c) i) deceleration = decrease in speed ÷ time ❶
　　　　　　= (15 m/s − 0 m/s) ÷ 2.4 s ❶
　　　　　　= 6.25 m/s² ❶
　　ii) force = mass × deceleration ❶
　　　　　= 850 kg × 6.25 m/s² ❶
　　　　　= 5 312.5 N ❶

5 a) i) **A** the motorcyclist is increasing
　　　　speed/accelerating ❶
　　　B the motorcyclist is travelling at a
　　　　constant speed ❶
　　　C the motorcyclist is increasing
　　　　speed/accelerating ❶
　　　D the motorcyclist is decreasing
　　　　speed/decelerating ❶
　　ii) acceleration = increase in velocity ÷ time
　　　taken ❶
　　　　　　= (8 m/s − 0 m/s) ÷ 10 s ❶
　　　　　　= 0.8 m/s² ❶
　　iii) force = mass × acceleration ❶
　　　　　= 220 kg × 0.8 m/s² ❶
　　　　　= 176 N ❶
　　iv) deceleration (negative acceleration)
　　　　= (14 m/s − 0 m/s) ÷ 20 s ❶
　　　　= 0.7 m/s² ❶
　　　Force = 220 kg × 0.7 m/s² ❶
　　　　　= 154 N ❶
　　v) The mass of the motorcycle is greater ❶
　　　Therefore the deceleration is less/it travels
　　　further during braking. ❶

Examiner's tip
▶ Avoid ambiguous phrases, such as 'It takes longer to stop', since it is not clear whether this refers to the distance or the time.

　　vi) Condition of tyres
　　　Condition of road
　　　Condition of brakes
　　　Tiredness of rider
　　　Whether medicine or alcohol has been
　　　consumed　　　any two, 1 mark each ❷

Examiner's tip
▶ Questions involving the use of **F = ma** are very common on higher tier papers. Parts iii) and iv) involve using the equation in calculations while part v) assesses your ability to use the equation to evaluate the effects of changing one factor.

6 a) Two arrows drawn that are equal in size ❶
　　One upwards and one downwards ❶
　　The upward force is the magnet's pull
　　on the paperclip ❶
　　The downward force is the Earth's
　　pull on the paperclip or the weight of the
　　paperclip ❶

Examiner's tip
▶ When describing forces, use the form 'object A pulls/pushes on object B'.

　b) The forces are equal in size ❶
　　Because the paperclip is not moving ❶

Examiner's tip
▶ The forces acting on any object that is not moving or is moving in a straight line at a constant speed are balanced.

　c) Paperclip is pulled to the left ❶
　　The attraction from the left magnet is
　　stronger ❶
　　Since it is closer to the paperclip ❶

Force and motion

To revise this topic more thoroughly, see Chapter 3 in *Letts Revise GCSE Physics Study Guide*.

 Try this sample GCSE question and then compare your answers with the Grade C and Grade A model answers on the next page.

Radar has a very short range in water, so ships use ultrasound to detect objects below the surface of the sea.

a   What is ultrasound?

.... [2]

b   A ship sends out a pulse of ultrasound and detects an echo 1.50 s later.

The speed of ultrasound in water is 1500 m/s.

  (i) Calculate the distance between the ship and the object that caused the echo.

....

.... [3]

  (ii) If the underwater object is moving at a speed of 10 m/s, how certain can the ship be of its position?

....

....

.... [3]

c   Ultrasound is also used to produce images of the inside of a human body.

It is used to scan the fetus of a woman who is pregnant.

  (i) Explain how ultrasound is used to scan a fetus.

....

....

.... [3]

  (ii) Why is ultrasound preferred to X-rays to produce an image of a fetus?

....

.... [2]

(Total 13 marks)

waves

GRADE C ANSWER

Alex gains one mark out of two here for her appreciation that ultrasound is above the range of human hearing. For the second mark she needed to make a statement about the type of wave, for example a longitudinal or compression wave.

No marks for a very vague answer. To gain credit here Alex needs to calculate the distance that the object could have moved while the ultrasound pulse was returning to the ship.

Another vague answer – this scores no marks.

Alex

a It's like sound, but it is too high a pitch for people to be able to hear it. ✓

b (i) Distance = speed × time ✓
= 1500 × 1.50 ✓ = 2250 m

(ii) 10 m/s isn't very fast, so they should know quite well.

c (i) The ultrasound bounces off the fetus so they can see where it is. ✓

(ii) Well, X-rays are very dangerous. They could kill the fetus and cause it to shrivel up.

Two marks out of three. This is a good answer, and well set out. The error Alex has made is in not realising that the ultrasound has travelled to the object and back again.

One mark for knowing that the fetus reflects ultrasound. For full marks more detail is needed about how the reflections are detected and used to produce an image.

4 marks = Grade C answer

Grade booster ⋯⋙ move a C to a B

Make sure that you have a detailed knowledge about the uses and dangers of different waves, particularly the waves that make up the electromagnetic spectrum. It is also worth remembering that when a distance is measured by the reflection of a wave, the wave has actually travelled twice the distance being measured.

GRADE A ANSWER

Full marks here – two out of two for a good description of ultrasound.

Two marks out of three for this answer. She has overlooked the fact that any uncertainty in the position of the object is due to the distance it moves while the sound is returning – this only takes 0.75 seconds.

No marks – more precise detail about the possible harmful effects is required.

Faiza

a It is a longitudinal wave like sound, but with a frequency that is too high for humans to detect. ✓✓

b (i) The ultrasound has travelled 500 × 1.5 = 2250 m, ✓ so the object is half this distance, i.e. 1125 m, ✓ because the sound has had to travel there and back. ✓

(ii) Not very certain, because in 1.5 s the object could have moved 15 m. ✓✓

c (i) First a jelly is spread over the woman's tummy. Then pulses of ultrasound are sent through this. Some of this ultrasound is reflected back. ✓

(ii) X-rays could harm the baby.

Full marks again – Faiza gains all three marks for this calculation – she even remembered to halve the distance travelled by the ultrasound.

Only one mark here for knowing that the fetus reflects the ultrasound.

8 marks = Grade A answer

Grade booster ⋯⋙ move A to A*

Be thorough in your preparation, and make sure that you know the uses and dangers of each type of electromagnetic wave. An A* candidate should have a good in-depth knowledge over a broad range of topics.

1 X-rays are used to examine bones where there is a suspected fracture.
The bone is placed between the X-ray source and a sheet of photographic film, as shown in the diagram.

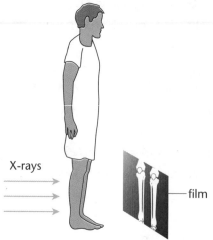

X-rays

film

a) Underline three phrases from the list that describe X-rays. ③

short wavelength	**long wavelength**	**electromagnetic waves**
low frequency	**high frequency**	**ultrasound waves**

b) Explain why bone appears white on the X-ray photograph and flesh appears dark.
....
.... ②

c) Explain how a doctor can tell whether the bone is fractured by studying the X-ray photograph.
....
.... ②

TOTAL 7

2

a) When light passes through a window it slows down as it enters the glass and speeds up again when it leaves. The speed of light in glass is about two-thirds of the speed of light in air. The diagrams show light waves about to travel through the glass. Complete the diagrams to show the passage of the light through the glass and out at the other side. ⑤

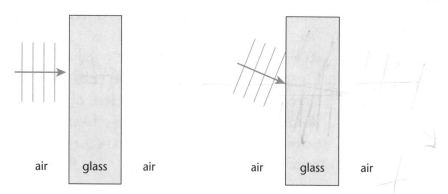

air   glass   air   air   glass   air

b) Periscopes use either mirrors or prisms to turn light round corners.
Complete the diagram of a periscope by showing how the light passes through the periscope and enters the eye. ④

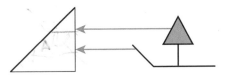

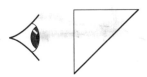

TOTAL 9

3 The diagram shows a transverse wave travelling along a rope.

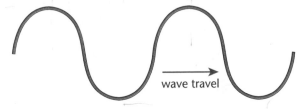

wave travel

a) Describe the movement of each part of the rope as the wave travels along it.

....

.... ②

b) Mark with an **A** a distance that is equal to the amplitude of the wave. ①

c) Mark with a **W** a distance that is equal to one wavelength of the wave. ①

d) When the frequency of the wave is 1.5 Hz it has a wavelength of 7.5 m. Calculate the speed of the wave along the rope.

....

....

.... ④

TOTAL 8

4 A person is sat in a room listening to music. The only lighting in the room is from a lamp in a ceiling fitting.

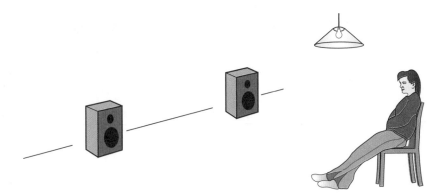

a) Explain how the person is able to see the loudspeakers.

....

.... ②

b) The loudspeakers are capable of reproducing sounds with frequencies up to 18 000 Hz. Sound travels in air at 330 m/s. Calculate the wavelength of a sound that has a frequency of 18 000 Hz.

....

.... ③

c) Loudspeakers are designed so that high frequency, short-wavelength sounds are emitted by a speaker with a small opening. Explain why loudspeakers are designed in this way.

....

....

.... ③

d) The diagram represents a person standing outside the room.

The sound from the loudspeakers reaches the person through the open doorway but the light does not.

  i) Use a diagram to explain how the sound reaches the person outside the room.　　③

  ii) Explain why the light from the lamp does not reach the person outside the room.

....

....

.... ③

TOTAL 14

5 The diagram shows light passing along an optical fibre.

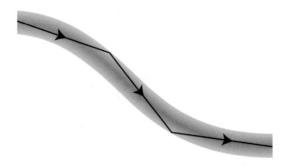

a) Explain how light is contained within the fibre.

....

....

.... ③

b) Information can be sent along optical fibres using either analogue or digital signals.

   i) The diagrams show an analogue and a digital signal.

     an analogue signal   a digital signal

   Describe the difference between these signals.

   ....

   .... ②

  ii) Signals that carry information need amplification.
     The diagrams compare the amplification of analogue and digital signals.

 an analogue signal   becomes noisy and distorted   which is amplified

 a digital signal   becomes noisy and distorted   and can be regenerated

   Use these diagrams to explain why the reception of a radio station from a digital signal is much clearer than that from an analogue signal.

   ....

   .... ②

c) An optical fibre communication system uses radiation that has a wavelength of 8.4×10^{-7} m.

   i) The speed of this radiation in the glass fibre is 2.2×10^{8} m/s.

     Calculate the frequency of the radiation.

     ....

     .... ③

  ii) Which part of the electromagnetic spectrum does this radiation belong to?

   .... ①

TOTAL 11

6 After an earthquake, two types of wave travel through the body of the Earth.
These are P waves (longitudinal waves) and S waves (transverse waves).
The diagram shows how these waves pass through the Earth.

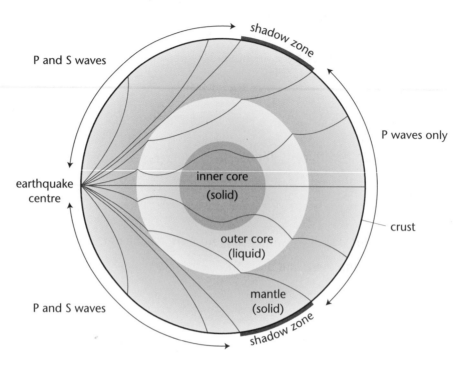

a) How does the diagram show that these waves change in speed as they pass through the Earth?

....

.... ②

b) Describe how the waves detected by a seismometer depend on its position on the Earth's surface.

....

....

.... ③

c) Explain how the detection of these waves gives evidence that part of the Earth's core is liquid.

....

....

....

.... ④

TOTAL 9

7 National radio and television programmes are transmitted from the Telecom tower. The information is carried by narrow beams of microwaves of wavelength 2.5 cm. After passing through a series of repeater stations where the signal is amplified, the information is broadcast on waves that have a wavelength of 60 cm. These waves can be detected by domestic television sets.

a) State one way in which the microwaves and those detected by a domestic television set:

  i) are similar

  .... ①

  ii) are different.

  .... ①

b) The diagram shows the dish aerial used to focus the microwaves into a narrow beam.

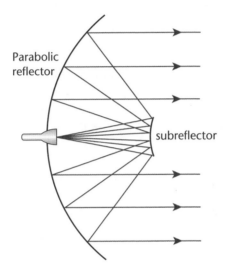

  i) Suggest why the microwaves are transmitted in a narrow beam.

  ....
  .... ②

  ii) Explain why a large diameter reflector has to be used to reduce the spreading of the beam.

  ....
  .... ②

  iii) Explain why the information is transmitted using microwaves rather than waves with a wavelength of 60 cm.

  ....
  ....
  .... ③

c) Microwaves can also be used to cook food.
  These microwaves have a wavelength of 11.6 cm.

  i) Why can these waves not be detected by a domestic television set?

  .... ①

  ii) How do microwaves cause food to become cooked?

  ....
  ....
  .... ③

TOTAL 13

QUESTION BANK ANSWERS

1 a) Short wavelength   ①
    High frequency   ①
    Electromagnetic waves   ①
b) X-rays do not pass through, or are absorbed by, bone   ①
    X-rays pass through, or are not absorbed by, flesh   ①
c) Bone appears white   ①
    A fracture shows up as black   ①

Examiner's tip
You should be familiar with the properties and uses of the waves that make up the electromagnetic spectrum.

2 a) The completed diagrams are shown below.

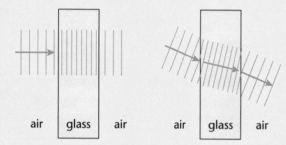

air   glass   air   air   glass   air

The marks are for:
Left-hand diagram
The wavelength in the glass is reduced   ①
The wavelength increases when the light emerges from the glass   ①
Right-hand diagram
The wavelength in the glass is reduced   ①
The change in direction as the light enters the glass is correct   ①
The waves leaving the glass are parallel to those entering   ①

Examiner's tip
A common error when drawing the right-hand diagram is to draw the waves in the glass parallel to the sides of the block.

b) One mark for each ray passing correctly through each prism.   ④
The correct diagram is:

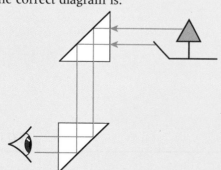

3 a) Each part of the rope vibrates or oscillates.   ①
    This movement is at right angles to the direction in which the wave is travelling.   ①
b) and c) The following diagram shows the amplitude (marked A) and the wavelength (marked W).

                1 mark each   ②

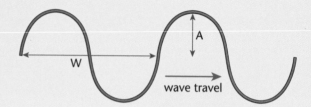

Examiner's tip
The amplitude is the distance from the centre position to the top of a peak or the bottom of a trough. One wavelength is the length of one complete cycle, i.e. a peak and a trough.

d) speed = frequency × wavelength   ①
      = 1.5 Hz × 7.5 m   ①
      = 11.25 m/s   ①
Quality of written communication mark – award if the answer is set out in a clear, logical way.   ①

4 a) The light comes from the lamp   ①
    It is reflected off the loudspeakers   ①

Examiner's tip
All objects that do not give off their own light are seen by the light that they reflect from the Sun and other light sources.

b) Wavelength = speed ÷ frequency   ①
             = 330 m/s ÷ 18 000 Hz   ①
             = 1.83×10^{-2} m   ①
c) The sound has to spread out from the loudspeaker   ①
Short wavelength waves need to pass through a narrow opening for sufficient spreading by diffraction.   ①
If a large opening were used the sound would not spread throughout the room.   ①

Examiner's tip
When answering questions on diffraction, you should always make a comparison between the wavelength and the size of the opening that the waves are being diffracted through. It is not enough to state that the opening is 'large' or 'small'.

FOR MORE INFORMATION ON THIS TOPIC ... SEE REVISE GCSE PHYSICS ... CHAPTER 3

d) i) Diffraction occurs at the doorway ❶
   This causes the sound to spread out as it
   passes through ❶

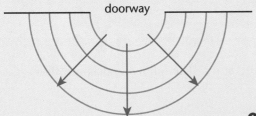

doorway

❶

   ii) The light does not spread out as it passes
       through the doorway ❶
       For light, the doorway is millions of
       wavelengths wide ❶
       So no diffraction occurs ❶

❺ a) Total internal reflection occurs at the
      boundary of the fibre ❶
      All the light is reflected, none passes
      through ❶
      This happens because the angle of incidence
      is greater than the critical angle ❶

Examiner's tip

*To gain full marks here, you need to give a
description and an explanation of total internal
reflection.*

   b) i) An analogue signal is continuously variable
         – it can take any value ❶
         A digital signal is either on or off ❶
      ii) When an analogue signal is amplified the
          noise is also amplified ❶
          Noise can be removed from a digital
          signal ❶

Examiner's tip

*Noise can be removed from a digital signal
because it should only have the values of 0 or
1. Anything in between or in excess of these
values is cleaned off.*

   c) i) frequency = speed ÷ wavelength ❶
         $= 2.2 \times 10^8 \, m/s \div 8.4 \times 10^{-7} m$ ❶
         $= 2.6 \times 10^{14}$ Hz ❶
      ii) infra red ❶

❻ a) The waves change direction ❶
      As they pass through different parts of the
      Earth and cross boundaries ❶

Examiner's tip

*The change in speed of the waves is called
refraction; it is the refraction or change in speed
that causes the change in direction.*

   b) A seismometer directly opposite the
      earthquake centre detects only the P
      waves ❶
      One close to the earthquake centre detects
      both types of wave ❶
      A seismometer in the shadow zone detects
      neither wave ❶

Examiner's tip

*You are not expected to be able to recall this
information. The question is testing your ability
to interpret the diagram.*

   c) Only P-waves are detected immediately
      opposite the earthquake centre ❶
      These must have travelled through a
      liquid ❶
      Since longitudinal waves can travel through a
      liquid ❶
      But transverse waves cannot ❶

Examiner's tip

*Transverse waves can travel across the surface of
a liquid, but not through the body of a liquid.*

❼ a) i) They are both transverse or electromagnetic
         or travel at the same speed in air/a
         vacuum ❶
      ii) They have different wavelengths or
          frequencies ❶
   b) i) So that the energy is not spread out ❶
         A narrow beam enables a smaller
         receiving dish to be used ❶
      ii) Diffraction occurs as the beam leaves the
          reflector ❶
          The greater the diameter of the reflector,
          the less spreading of the beam occurs ❶

Examiner's tip

*On Higher tier physics papers questions about
diffraction are common. You need to be able to
recognise situations where the spreading out of
waves is due to them having a wavelength
comparable to the size of the opening that they
are passing through.*

      iii) Waves with a wavelength of 60 cm would
           need very large reflectors ❶
           To minimise the effects of diffraction ❶
           The diameter of the reflector needs to be
           many times the wavelength of the wave ❶

   c) i) The wavelength is too short to be
         detected ❶
      ii) The energy is absorbed by the water
          molecules in the food ❶
          This increases the temperature of the
          water ❶
          Energy is transferred to other food
          particles by conduction ❶

waves

The Earth and beyond

To revise this topic more thoroughly, see Chapter 4 in *Letts Revise GCSE Physics Study Guide.*

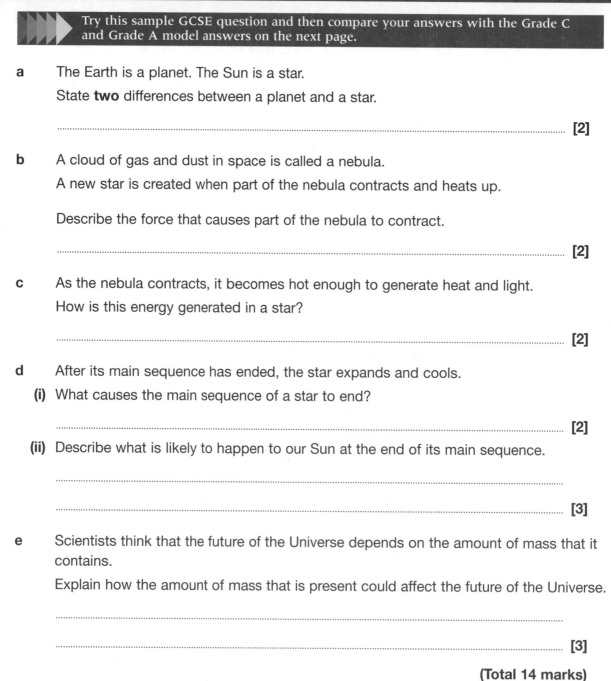

Try this sample GCSE question and then compare your answers with the Grade C and Grade A model answers on the next page.

a   The Earth is a planet. The Sun is a star.
State **two** differences between a planet and a star.

.... [2]

b   A cloud of gas and dust in space is called a nebula.
A new star is created when part of the nebula contracts and heats up.

Describe the force that causes part of the nebula to contract.

.... [2]

c   As the nebula contracts, it becomes hot enough to generate heat and light.
How is this energy generated in a star?

.... [2]

d   After its main sequence has ended, the star expands and cools.
  (i) What causes the main sequence of a star to end?

.... [2]

  (ii) Describe what is likely to happen to our Sun at the end of its main sequence.

....

.... [3]

e   Scientists think that the future of the Universe depends on the amount of mass that it contains.

Explain how the amount of mass that is present could affect the future of the Universe.

....

.... [3]

(Total 14 marks)

These two answers are at Grade C and A. Compare which one your answer is closest to and think how you could have improved it.

GRADE C ANSWER

Full marks are awarded to Seth for this answer, which gives two correct differences between a planet and a star.

No marks. This is a common misunderstanding by grade C candidates at GCSE, they think that stars are like giant bonfires, whereas no burning reaction takes place.

In his first sentence, Seth only repeats what has been given in the question. Again, no marks.

Seth

a Planets are not hot enough to give out light, which stars do. Also, planets go round stars and stars do not go round planets. ✓✓

b All the bits of gas and dust pull together. ✓

c The gases burn in an enormous fire. ✗

d (i) It runs out of gas. ✗

  (ii) It will expand and cool. Eventually it will burn right out. ✗

e The more mass there is, the slower it will move. So a more massive Universe will take longer to expand. ✗

3 marks = Grade C answer

Just one mark is awarded here for realising that the forces are attractive. The second mark would be for stating that the forces are gravitational.

Seth is on the right lines here, but unfortunately he's still on the idea of burning, so no marks.

Another common misunderstanding that the more mass, the slower things move. The Sun and Solar System whirl round the centre of our galaxy at enormous speed.

Grade booster ⋯⟩ move a C to a B
A grade B candidate should show a better understanding of how a star generates energy by nuclear fusion and about the stages in the life cycle of a star.

GRADE A ANSWER

Two marks out of two. Rashid's answers are clear, precise and expressed very well.

One mark out of three. The question is asking candidates to describe what happens after the expansion to a red giant. Rashid has stopped short – she should have stated that the star will then contract to a white dwarf, followed by cooling to form a black dwarf.

Both of these statements are correct, but gain no marks. The possible future of the Universe needs to be linked to the amount of mass present. There are in fact three possibilities: continued expansion, contraction or reaching a steady state.

Rashid

a Planets orbit stars (not the other way round) and stars give out light, but planets only reflect light from stars. ✓✓

b This is the gravitational attractive force between the particles in the nebula. ✓✓

c It is the joining together of small nuclei to make larger ones. For example, the fusion of hydrogen nuclei to form helium. This process is called fission and it releases energy. ✓

d (i) All the gas is used up. ✗

  (ii) It will become a red giant before contracting again. ✓

e It could carry on expanding for ever, or it could contract.

6 marks = Grade A answer

Full marks again here for correct answer, clearly expressed.

One mark out of two is awarded for this response. Rashid's error is in confusing the processes of fission, which is about the splitting of large nuclei, with fusion, the reaction that takes place in a star. She has described this reaction correctly, which is why she is awarded one mark.

No marks here. The main sequence ends when there are not enough hydrogen nuclei left to sustain it.

Grade booster ⋯⟩ move A to A*
An A* candidate should have a good understanding of how fusion is the source of energy in the Universe and to give a reasoned argument about how the future of the Universe depends on the amount of mass that it contains.

The Earth and beyond

QUESTION BANK

1 There are thousands of artificial satellites in orbit around the Earth.

a) Give **two** uses of artificial satellites.

....

.... ②

b) Some satellites orbit the Earth above the equator with an orbit time of 24 hours. These are called geostationary satellites.
Explain why 24 hours is a suitable orbit time for some satellites.

....

....

.... ③

c) The graph shows how the orbit time of a satellite depends on its height above the Earth's surface.

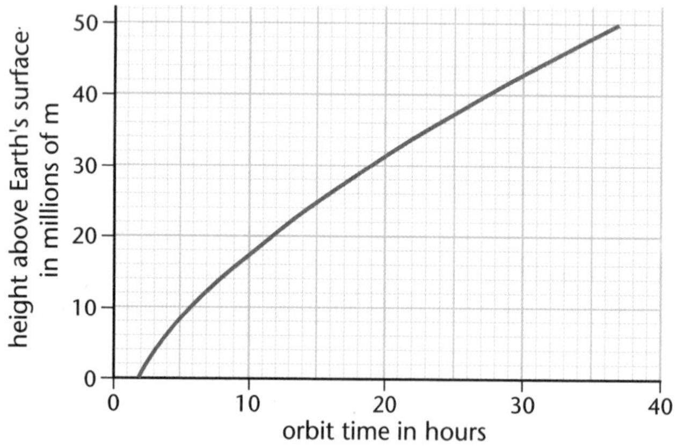

i) Describe how the orbit time of a satellite depends on its height above the Earth's surface.

....

.... ②

ii) Use the graph to write down the height above the Earth's surface of a geostationary satellite.

.... ①

d) Some satellites are in low orbits around the North and South Poles. They complete one orbit in 90 minutes.

i) Explain how the force acting on a satellite in a low polar orbit compares to the force on a satellite in a high equatorial orbit.

....

.... ②

ii) Describe the advantages of a low polar orbit for some satellites.

....

....

.... ③

TOTAL 13

2 The table gives some information about each of the five outer planets.

Planet	Radius in km	Mass (Earth = 1)	Density in g per cm³	Surface gravitational field strength in N/kg	Surface temperature in °C	Radius of orbit (Earth = 1)	Orbital speed in km/s
Jupiter	71 000	318	1.33	23	−150	5.2	13.1
Saturn	60 000	95	0.71	8.9	−180	9.5	9.6
Uranus	25 500	15	1.24	8.7	−215	19.1	6.8
Neptune	24 800	17	1.67	11	−220	30.0	
Pluto	1 120	0.025	1.95	0.72	−225	39.5	4.7

a) Describe the relationship between the surface temperature of a planet and its distance from the Sun.

.... ①

b) Which **two** planets are similar in size and mass?

....

.... ①

c) Which two planets are the closest together?

.... ①

d) Explain why Neptune is more massive than Uranus, even though it is smaller.

.... ①

e) If it were possible to land an astronaut on each planet, on which planet would the astronaut find it very difficult to jump? Give the reason for your answer.

....

.... ②

f) Use the grid to draw a graph of orbital speed (*y*-axis) against radius of orbit (*x*-axis).　④

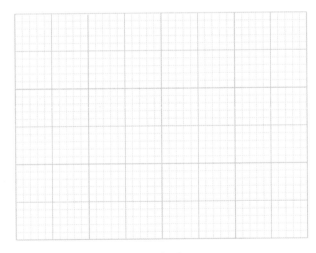

g) Use your graph to deduce the orbital speed of Neptune.

.... ①

TOTAL 11

3 Our Sun is a small star in its main sequence.

a) Describe the reaction that takes place in the core of the Sun.

....

.... ②

b) It is likely that at the end of its main sequence the Sun will expand.
Describe how the colour and temperature of the Sun are likely to change as it expands.

....

.... ②

c) After the expansion, the Sun is likely to contract again.

i) What causes a sun to contract?

.... ①

ii) Describe what is likely to happen to the Sun after it has contracted.

....

.... ②

TOTAL 7

4 Comets, like the planets, go round the Sun in elliptical orbits. The orbit times of comets range from a few years to millions of years. The diagram shows a comet orbit.

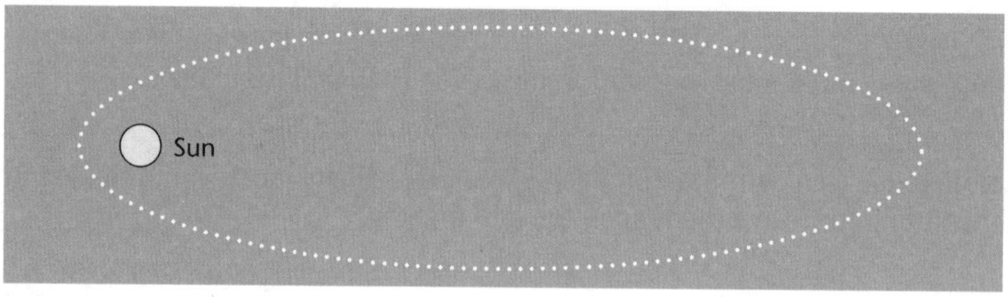

a) Describe and explain how the speed of a comet changes as it approaches the Sun.

....

.... ②

b) Mark an 'S' on the diagram at the point where you would expect the comet to have its smallest speed. ①

c) Comets have a structure similar to that of the moons of the outer planets, mainly ice. Comets become visible as a glow when their distance from the Sun is approximately three times the radius of the Earth's orbit.

i) Suggest what is happening to a glowing comet to make it visible.

....

.... ②

ii) Describe how you would expect the mass of a comet to change as it completes a single orbit around the Sun.

....

.... ②

d) Comets leave trails of dust in their wake as they pass close to the Sun. Bright lights can be seen in the Earth's atmosphere when the Earth passes through these dust clouds. Suggest what happens to cause these bright lights.

....

.... ②

TOTAL 9

5 Astronomers can measure the speed at which galaxies are moving away from the Milky Way.

a) Explain how astronomers measure the speed of a galaxy relative to the Milky Way.

....

.... ②

b) How does the speed of a galaxy relative to the Milky Way depend on its distance from the Milky Way?

.... ①

c) Explain how changes in the rate at which the Universe is expanding could give astronomers a clue to its future.

....

....

.... ④

TOTAL 7

The Earth and beyond

❶ a) Monitoring the weather
Spying
Communications
Navigation   any two, 1 mark each **❷**
b) It is the same time as it takes the Earth to
turn on its axis **❶**
So the satellite remains above the same point
on the Earth's surface **❶**
This is needed for communications
satellites **❶**
c) i) The orbit time increases **❶**
With increasing height above the Earth's
surface **❶**
ii) 36 million metres **❶**

Examiner's tip

*These questions are testing your skill in
handling data; in this case at interpreting data
in the form of graphs. Data handling is an
important scientific skill that is tested at GCSE
level.*

d) i) The force on the low orbit satellite is
greater (for the same mass) **❶**
Because gravitational forces decrease
with increasing separation **❶**
ii) The satellite completes a number of
orbits (16) each day **❶**
It sees a slightly different view of the
Earth on each orbit **❶**
This is useful for monitoring the weather
over the whole of the Earth's surface **❶**

❷ a) The further a planet is from the Sun, the
colder its surface **❶**
b) Uranus and Neptune **❶**
c) Jupiter and Saturn **❶**
d) Neptune is denser than Uranus **❶**
e) Jupiter **❶**
The gravitational field strength is greatest
on Jupiter, so the astronaut would be very
heavy. **❶**

Examiner's tip

*The gravitational field strength measures the
size of the pull on each kilogram of material. A
60 kg person who weighs 600 N on Earth
would weigh 1380 N on Jupiter.*

f) Here is the completed graph:

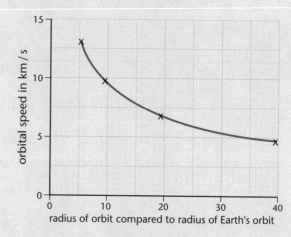

Marks are awarded for:
Correct scales and labels on axes **❶**
Plotting all four points correctly **❷**
One mark for plotting two or three correctly.
Drawing a smooth curve (not dot-to-dot) **❶**
g) Answer in the range 5.2 – 5.5 km/s **❶**

Examiner's tip

*This question is concerned with the scientific
skills of evaluation and data handling. These
are important skills that are assessed on all
physics examination papers and practice of
these skills should form an important part of
your revision. It is important that you do not
restrict your revision to learning facts, recall of
factual information only represents about 25%
of the marks in physics examinations.*

❸ a) It is a fusion reaction **❶**
Hydrogen nuclei join together to form
helium nuclei **❶**
b) The Sun will change colour to red **❶**
Its temperature will drop **❶**

Examiner's tip

*Expansion causes cooling and contraction
causes heating.*

c) i) Gravitational force that pulls all the
      particles together   **1**
   ii) It will cool   **1**
      Its colour will change/its brightness
      will fade   **1**

Examiner's tip

▶▶▶ *If you described the changes that take place as
the Sun contracts, you should take more care to
read what the question is asking.*

❹ a) The comet speeds up as it approaches the
      Sun   **1**
      This is caused by the increasing gravitational
      pull   **1**
   b) The 'S' should be at the extreme right-hand
      side of the comet's orbit on the diagram   **1**

Examiner's tip

▶▶▶ *All the time that the comet is travelling away
from the Sun, it is slowing down because the
gravitational force is pulling it in the
'backwards' direction. It starts to speed up
when it changes direction and heads towards
the Sun again.*

   c) i) Water vapour evaporates from the ice   **1**
         This then reflects light from the Sun   **1**
     ii) The mass decreases when the comet is
         near the Sun   **1**
         The mass stays the same when the comet
         is distant from the Sun   **1**
   d) The dust enters the Earth's atmosphere   **1**
      Where it burns up as it falls towards the
      Earth   **1**

❺ a) They measure the change in frequency
      or wavelength of the light emitted   **1**
      The greater the 'red shift', the faster the
      galaxy is moving   **1**

Examiner's tip

▶▶▶ *The wavelength and frequency of the light that
we detect from a galaxy are not the same as
those that are emitted. The wavelength is
increased and the frequency decreased due to
the motion of a galaxy moving away from the
Milky Way.*

   b) The further a galaxy is from the Milky
      Way, the faster it is moving.   **1**
   c) An increasing or constant rate of expansion
      indicates that the Universe could go on
      expanding for ever.   **1**
      A decreasing rate of expansion indicates
      that the Universe could reach a stable
      (unchanging) size   **1**
      Or it could start to collapse.   **1**
              Quality of written communication
              mark – the information is relevant
                          to the question. **1**

The Earth and beyond

Energy

To revise this topic more thoroughly, see Chapter 5 in *Letts Revise GCSE Physics Study Guide*.

 Try this sample GCSE question and then compare your answers with the Grade C and Grade A model answers on page 52.

The diagram shows a crane.

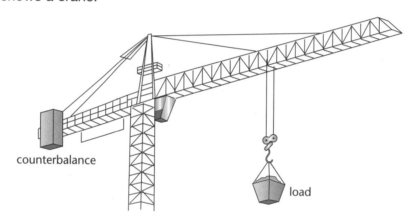

counterbalance

load

a   What is the purpose of the counterbalance?

    ....

    .... **[2]**

b   The crane is lifting a load of 80 000 N.

  (i) The crane lifts the load through a vertical distance of 15.0 m.

    Calculate the potential energy gained by the load.

    ....

    ....

    .... **[3]**

  (ii) The load is lifted at a speed of 0.3 m/s.

    Calculate the time it takes for the crane to lift the load 15.0 m.

    ....

    ....

    .... **[3]**

(iii) Use your answers to **(i)** and **(ii)** to calculate the power output of the crane in lifting the load.

....

....

.... **[3]**

c The power input to the crane is 72 kW.

Calculate the efficiency of the crane.

....

....

.... **[2]**

(Total 13 marks)

These two answers are at Grade C and A. Compare which one your answer is closest to and think how you could have improved it.

GRADE C ANSWER

Mark does not gain any marks for this answer. He has not appreciated that this part of the question is about the turning effect of forces and that the counterbalance is needed to exert a moment in opposition to the moment of the load.

Full marks for recall and correct use of this relationship.

Now Mark is totally off the track. He has remembered something about power, but this is the wrong context for this relationship. No marks.

Mark

a   It's there to stop the crane from falling over. It counteracts the weight. ✗

b (i) Potential energy = mass × g × height ✓ = 1500 × 10 × 15
= 225 000 J ✗

(ii) Time = distance ÷ speed ✓
= 15 ÷ 0.3 = 50 s ✓✓

(iii) Power = voltage × current
= 1500 × 15 = 22 500

c   Efficiency = input ÷ output
= 72 ÷ 22 500 = 0.0032 ✗

Like many GCSE candidates at this level, Mark shows that he is confused between the concepts of mass and weight. He is awarded one mark for quoting the correct relationship, but he is unable to use it correctly.

This relationship is given in the question paper or data book. Unfortunately Mark has miscopied it. When a candidate remembers or copies a relationship wrongly, no marks at all are awarded.

3 marks = Grade C answer

Grade booster ····⟩ move a C to a B
This question involves a number of calculations and Mark is not confident about carrying any of them out. At grade B a candidate is expected to have sorted out the meanings of mass and weight and the relationship between them. It is also expected that a grade B candidate will be able to quote correct units for physical quantities.

GRADE A ANSWER

Full marks for a correct answer. Gita has appreciated that the load exerts a turning effect on the crane and that a moment in the opposite direction is needed to prevent it from toppling.

One mark out of two here. This relationship is given to candidates, so there is no mark available for recalling it. She gains a mark for a correct substitution, even though the 2.4 kW is a wrong answer, but she thinks that she has calculated a percentage efficiency. To express the efficiency as a percentage, she would need to multiply her answer by 100.

Gita

a   It stops the crane from toppling by providing a turning effect that balances the turning effect of the load. ✓✓

b (i) Potential energy = weight × change in height ✓ = 80 000 N × 15 m
= 1.2 × 10⁶ J ✓✓

(ii) Time = distance ÷ speed ✓
= 15.0 m ÷ 0.3 m/s = 50 s. ✓✓

(iii) Power = energy transfer ÷ time ✓
= 1.2 × 10⁶ J ÷ 50 s ✓ = 2.4 kW ✗

c   efficiency = power output ÷ power input = 2.4 kW ÷ 72 kW ✓ = 0.033%

Full marks again for a correct calculation and unit.

Not surprisingly, Gita completes this calculation correctly and again is awarded full marks.

In this part of the question Gita has recalled the relationship correctly and substituted the correct values for the physical quantities. Unfortunately she has tried to convert the answer in W to one in kW and made an error – she is out by a factor of 10. So, only two marks out of three here.

11 marks = Grade A answer

Grade booster ····⟩ move A to A*
Gita's response to this question is very good, but she does not show the confidence with physical quantities and units that is expected of an A* candidate. To qualify for an A* grade, a candidate should show no hesitation in converting from W to kW or knowing how to express efficiency as a fraction or as a percentage.

Energy

1 The generator in a power station consists of an electromagnet that rotates inside three sets of copper conductors.

a) Explain why a current passes in the conductors when the electromagnet rotates.

....

.... ②

b) The electricity passes from the generator to a transformer where the voltage is increased to 400 000 V before it passes into the national grid.

  i) The primary coil of the transformer has 5000 turns. Calculate the number of turns on the secondary coil.

  ....

  ....

  .... ③

  ii) Explain why the voltage is increased before the electricity is transmitted.

  ....

  ....

  .... ③

c) A coal-burning power station has an efficiency of 40%.
   Describe what happens to the energy released from the burning coal.

  ....

  ....

  .... ③

TOTAL 11

2 Solar power is unreliable in Britain, but in some countries many houses use energy from the Sun to produce hot water for domestic use. The diagram shows how this can be done.

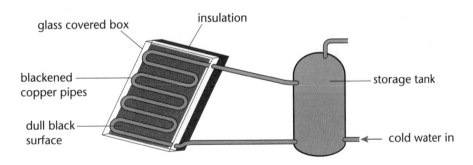

a) Explain why the pipes are painted black.

.... ①

b) State **two** reasons why copper pipes are used rather than iron pipes.

....

.... ②

c) Explain why the pipes are contained in a glass-covered box.

....

.... ②

d) Describe and explain how the water circulates in the system.

....

....

.... ③

TOTAL 8

❸ In some exposed parts of Britain the wind blows all the time. 'Wind farms' consisting of many wind turbines can produce as much electricity as a small coal-fired power station. A typical wind turbine has an electrical power output of 3 MW ($1 \text{ MW} = 1 \times 10^6 \text{ W}$).

a) The power of the wind blowing through the turbine is 9 MW. Use the equation:

efficiency = useful power output ÷ total power input

to calculate the efficiency of the turbine when the power output is 3 MW.

....

.... ②

b) One disadvantage of wind turbines is the high cost of manufacture and installation. State **three** advantages of wind turbines over a coal-fired power station.

....

....

.... ④

c) State **two** other disadvantages of using wind turbines to generate electricity.

....

.... ②

TOTAL 8

❹ A cycle and cyclist have a combined mass of 80 kg.

a) Calculate the kinetic energy of the cycle and cyclist when travelling at a speed of 12 m/s.

....

....

.... ③

b) The cyclist applies a constant braking force and comes to rest after 8.0 s.

  i) Calculate the power of the brakes.

....

....

.... ③

  ii) The braking distance is 48 m. Calculate the size of the braking force required.

....

....

.... ③

  iii) Calculate the kinetic energy of the cycle and cyclist when travelling at a speed of 6 m/s.

....

.... ②

  iv) Explain why, when the same braking force is applied, the braking distance from a speed of 6 m/s is 12 m.

.... ①

TOTAL 12

5 The diagram shows a magnet and a coil of wire which is connected to a sensitive ammeter. When the magnet is moved slowly into the coil the needle on the ammeter shows a deflection to the right.

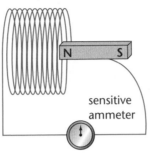

sensitive
ammeter

a) Explain why there is a reading on the ammeter when the magnet is moved into the coil.

....

.... ②

b) Describe and explain the ammeter reading when the magnet is:

i) held steady inside the coil

....

.... ②

ii) withdrawn slowly from the coil

....

.... ②

iii) moved quickly in and out of the coil.

....

.... ②

TOTAL 8

---

6 The generators in a power station produce electricity at 25 000 V. This is increased to 400 000 V before passing into the National Grid.

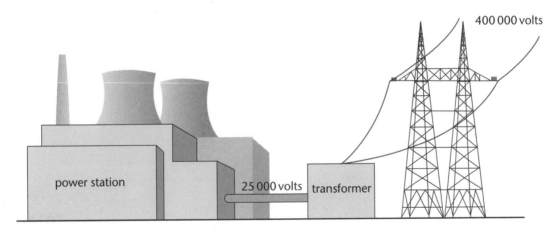

400 000 volts

power station   25 000 volts   transformer

a) Explain why electricity is transmitted at a high voltage.

....

....

.... ③

b) The primary coil of the transformer has 10 000 turns. Calculate the number of turns on the secondary coil.

....

.... ③

c) The high voltage electricity travels along overhead cables supported on pylons.
It is possible to use underground cables instead.
Suggest **two** advantages and **two** disadvantages of underground cables.

....

....

....

.... ④

TOTAL 10

7 The diagram shows how a house can be insulated by filling the cavity between the inner and outer walls with mineral wool.

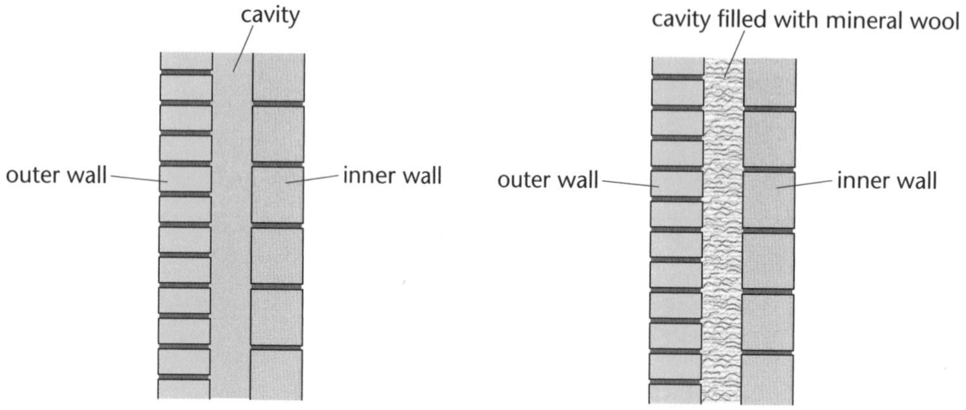

a) What is the purpose of insulating a house?

....

.... ②

b) How is energy transferred through an uninsulated cavity between the inner and outer walls of a house?

....

.... ②

c) Explain how the mineral wool used to insulate a cavity provides effective insulation.

....

....

.... ③

d) Explain how insulating houses benefits the environment.

....

....

.... ③

TOTAL 10

8 The diagram shows a d.c. motor.

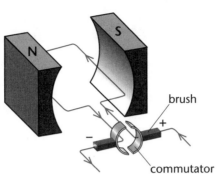

brush

commutator

a) i) Draw an arrow to show the direction of the magnetic field between the poles of the magnet. ①

ii) Explain why the forces acting on the sides of the coil are in opposite directions and describe the directions of the forces.

....

.... ②

iii) Explain why there is no turning effect when the plane of the coil is vertical.

....

.... ②

b) What is the purpose of the commutator? Explain how it does this.

....

....

.... ③

TOTAL 8

Energy

1 a) The changing magnetic field ❶
   Induces or causes a current ❶

Examiner's tip
Always emphasise the change in the magnetic field when explaining the effects of electromagnetic induction.

b) i) $\dfrac{\text{primary turns}}{\text{secondary turns}} = \dfrac{\text{primary voltage}}{\text{secondary voltage}}$ ❶

$\text{secondary turns} = \dfrac{\text{primary turns} \times \text{secondary voltage}}{\text{secondary turns}}$ ❶

$\dfrac{5000 \times 400\,000}{25\,000} = 80\,000$ ❶

Examiner's tip
It is usually easier to use ratios when calculating transformer turns or voltages. In this case the ratio of the voltages is 1 : 16, so the number of turns must be in the same ratio.

ii) A low current is needed for transmission ❶
   A high current causes excessive heating of the cables ❶
   This results in power loss ❶

c) 40% goes to electricity ❶
   60% is wasted ❶
   The wasted energy ends up as heat in the surroundings ❶

2 a) Black is the best absorber of radiant energy ❶

b) Copper is a better conductor of thermal energy (heat) ❶
   Copper does not corrode ❶

c) The box acts like a greenhouse ❶
   It reduces the amount of energy lost by radiation ❶

Examiner's tip
An alternative answer would be that the glass covering allows radiant energy to enter, but does not allow air to enter and cool the pipes.

d) Water is heated in the pipes and rises ❶
   This happens because it is less dense than the colder water ❶
   Cold, denser water from the bottom of the tank moves down to replace it ❶

3 a) Efficiency = 3 MW ÷ 9 MW ❶
   = 0.33 ❶

Examiner's tip
Efficiency is expressed as a number less than 1 or as a percentage. It does not have a unit.

b) Wind turbines have low running costs ❶
   They do not use fossil reserves ❶
   They do not cause atmospheric pollution ❶
   Quality of written communication mark – the answer is clear and unambiguous. ❶

c) Wind turbines occupy a lot of room ❶
   They are noisy/do not work if there is no wind ❶

4 a) Kinetic energy = $\frac{1}{2} \times m \times v^2$ ❶
   = $\frac{1}{2} \times 80$ kg $\times (12 \text{ m/s})^2$ ❶
   = 5760 J ❶

Examiner's tip
When working out kinetic energy using a calculator, first square the speed, then multiply this by half the mass.

b) i) Power = energy transfer ÷ time taken ❶
   = 5760 J ÷ 8 s ❶
   = 720 W ❶

Examiner's tip
You must have the correct answer and unit for the third mark. Note that although a watt is equivalent to a joule/second, J/s is not usually accepted as the unit of power.

ii) Energy transfer = force × distance moved ❶
   Force = energy transfer ÷ distance moved
   = 5760 J ÷ 48 m ❶
   = 120 N ❶

Examiner's tip
An alternative approach to answering this question is to calculate the acceleration as – 1.5 m/s² and then to use F = ma. This would be awarded full marks in an examination.

iii) Kinetic energy = $\frac{1}{2} \times m \times v^2$
   = $\frac{1}{2} \times 80$ kg $\times (6 \text{ m/s})^2$ ❶
   = 1440 J ❶

iv) There is only one quarter of the energy to be removed from the cycle and cyclist, which is achieved by the force working for one quarter of the distance ❶

5 a) There is a changing magnetic field in and around the coil ❶
This induces a voltage which causes a current in the coil ❶

Examiner's tip

Notice how this answer emphasises that the changing magnetic field causes the induced voltage.

b) i) There is no reading on the ammeter ❶
Because the magnetic field around the coil is not changing ❶
ii) The ammeter deflection is to the left ❶
Because the change of field has been reversed ❶

Examiner's tip

Note that reversing the magnet or the direction of movement causes the current to pass in the opposite direction.

iii) The current alternates, changing direction when the movement of the magnet changes direction ❶
The current is bigger than before because the movement is faster, so the magnetic field changes at a greater rate. ❶

6 a) So that power is transmitted at a low current ❶
This reduces energy losses ❶
Due to heating in the transmission wires ❶

b) $$\frac{\text{primary turns}}{\text{secondary turns}} = \frac{\text{primary voltage}}{\text{secondary voltage}}$$ ❶

$$\text{secondary turns} = \frac{\text{primary turns} \times \text{secondary voltage}}{\text{secondary turns}}$$ ❶

$$\frac{10\,000 \times 400\,000}{25\,000} = 160\,000$$ ❶

Examiner's tip

If you worked out the ratio of the voltages, your answer should be 1 : 16, so the number of turns on the secondary coil is 16 times that on the primary coil.

c) Advantages
They are not unsightly
They are less easily damaged in bad weather
They do not pose a danger to people or buildings   any two, 1 mark each ❷
Disadvantages
The cables are expensive to manufacture
They are expensive to install
Running costs are high because of the need for a coolant   any two, 1 mark each ❷

7 a) To reduce the heat loss ❶
From the warm interior to the cold exterior ❶

Examiner's tip

Insulation reduces the energy transfer between objects at different temperatures. In summer, domestic insulation helps to keep a house cool but the purpose of installing it is to keep a house warm and reduce heating bills in spring, autumn and winter.

b) It travels through the solid walls by conduction ❶
It travels through the cavity mainly by convection currents ❶

Examiner's tip

There is some energy transfer by conduction across an uninsulated cavity, but the amount is small compared to that which is transferred by convection.

c) The mineral wool traps pockets of air ❶
The air is unable to form convection currents ❶
It can only conduct energy through the cavity, and air is a poor thermal conductor ❶
d) There is less demand for fossil fuels such as oil and gas to burn in central heating boilers ❶
Less carbon dioxide, which increases the greenhouse effect, is produced ❶
Less sulphur dioxide, which causes acid rain, is produced ❶

8 a) i) The arrow should go from left to right ❶

Examiner's tip

The direction of a magnetic field is the direction of the force that would be exerted on the N-seeking pole of a magnet; i.e. from N-seeking to S-seeking poles.

ii) The currents are in opposite directions so the forces are in opposite directions ❶
The forces are vertical ❶
iii) The forces on the sides of the coil are parallel to its plane ❶
There is no moment (turning effect) as the perpendicular distance from the force lines to the axis of rotation is zero ❶
b) The commutator is there to keep the coil rotating in the same direction (clockwise or anticlockwise) ❶
It does this by reversing the direction of the current in the coil ❶
Once each half-revolution ❶

Examiner's tip

Candidates often confuse the purpose of the commutator with that of the brushes. The brushes pass current to and from the coil. Without the commutator, the motor could not rotate.

CHAPTER 6

Radioactivity

To revise this topic more thoroughly, see Chapter 6 in *Letts Revise GCSE Physics Study Guide.*

 Try this sample GCSE question and then compare your answers with the Grade C and Grade A model answers on pages 61 and 62.

Radon-222 is unstable. It decays by emitting an alpha particle.

a (i) Describe the structure of an alpha particle.

....

.... **[2]**

(ii) From which part of the atom is the alpha particle emitted?

.... **[1]**

(iii) Following the emission of alpha radiation, an atom is often left with excess energy.
It gets rid of this by emitting high-energy radiation that has no mass or charge.
What type of radiation is emitted?

.... **[1]**

b A Geiger–Müller tube and counter are used to measure the activity of a sample of radon-222. The result is 120 counts/s.

(i) Describe how this activity is measured.

....

....

.... **[4]**

(ii) Explain why the sample of radon-222 is likely to be decaying at a greater rate than this.

.... **[1]**

(iii) The half-life of radon-222 is 4 days. Eight days after the initial reading, the activity is measured again. Estimate the result of this measurement.

....

....

.... **[3]**

(iv) Explain why your answer to **b (iii)** can only be an estimate.

.... **[1]**

(Total 13 marks)

GRADE C ANSWER

Rana is awarded both of the available marks for stating that an alpha particle consists of two neutrons and two protons.

Rana has only repeated the information given in the question. She gains no marks. The correct answer is gamma radiation.

Rana just scrapes a mark here. The radiation is given off in all directions, so the G–M tube only detects a fraction of it.

No marks again. Unlike chemical reactions, radioactive decay is not affected by physical conditions such as temperature.

Rana

a (i) It has two of each, protons and neutrons. But no electrons. It's a bit like helium really. ✓✓

(ii) From the middle bit, the nuclide.

(iii) Very high-energy radiation. This radiation has no mass or charge. ✗

b (i) Point the tube at the sample and then turn it on. After 1 second turn it off. Then do it again and take an average. ✓

(ii) The G-M tube does not detect all the radiation. Some will go out in the opposite direction. ✓

(iii) If 4 days is the half-life, then 8 days is the dead life, so there is none left. ✗

(iv) If it was cold it might decay slower, and if it was hot it might decay quicker. ✗

Wrong scientific terminology costs Rana the mark here. She should have written 'nucleus'.

Only one mark here. One second is too short a time interval and she does not appreciate that background radiation needs to be taken into account. However, she gains her mark for stating that a repeat reading needs to be taken.

No marks. Rana does not understand the concept of half-life.

4 marks = Grade C answer

Grade booster ····∴ move a C to a B
This question involves a number of high-level concepts and Rana is unsure about many of them. To move to a grade B, she needs to be confident with calculations involving half-lives and to have some understanding about the random nature of radioactive decay.

Radioactivity

GRADE A ANSWER

Tim

Totally correct – two marks. Notice how Tim does not pad out his answers with irrelevant information – this shows that he is confident about his understanding of this topic.

This is a very good response. Three marks out of four. The only thing that is missing is the need to repeat the measurement. This is necessary because radioactive decay is a random process, and measurements at different times will give different results.

Correct – all three marks are awarded.

No marks. Again, Tim loses marks because he does not understand the effects of the random decay of radioactive nuclei.

a (i) An alpha particle is two neutrons and two protons – identical to a helium nucleus. ✓✓

(ii) It comes from the nucleus. ✓

(iii) This will be X-rays, because this is a high-energy radiation with no mass or charge. ✗

b (i) First you need to measure how many counts in one minute, say. Then point the tube away from the radon to measure the background count and take this away. Finally divide by the number of seconds in a minute, 60. ✓✓✓

(ii) There must be some that is not getting through to the detector. ✗

(iii) 8 days is two half-lives. ✓
So the activity is $\frac{1}{2} \times \frac{1}{2} \times 120$ counts/s = 30 counts/s. ✓✓

(iv) I might have the wrong answer. ✗

Correct – one mark.

No marks here because Tim does not know the difference between X-rays and gamma rays. The waves are the same, but come from different sources. X-rays come from X-ray machines. Gamma rays are emitted by a nucleus that has too much energy to be stable.

Not enough for a mark here. Tim needs to state what is happening to the radiation that does not reach the detector.

9 marks = Grade A answer

Grade booster ⋯⟩ move A to A*

Tim's overall response is very good, and there are few weaknesses. He falls down on not knowing that X-rays and gamma rays differ in their origin, and in not appreciating that radioactive decay is a random process, so results are not precisely predictable.

1

a) Here are three descriptions of the main types of nuclear radiation.
   A Short wavelength electromagnetic radiation.
   B A particle consisting of two protons and two neutrons.
   C A fast-moving electron.
   Which lettered statement is a description of:

   i) an alpha particle .... ①

   ii) a beta particle .... ①

   iii) gamma radiation? .... ①

   Radon-220 is a radioactive gas that seeps from underground rocks. It decays by emitting alpha particles and it has a half-life of 52 s.

b) Explain the meaning of the term half-life.

   ....

   .... ②

c) The activity of a sample of radon-220 is measured to be 520 counts/s.

   i) Name a suitable instrument for measuring the activity of the radon-220.

   .... ①

   ii) Estimate the activity of the radon-220 after

   52 s .... 104 s .... ②

   iii) Explain why it is not possible to make a precise prediction of the activity of the radon gas.

   ....

   .... ②

   TOTAL 10

2 The diagram compares the levels of background radiation in some parts of the UK.

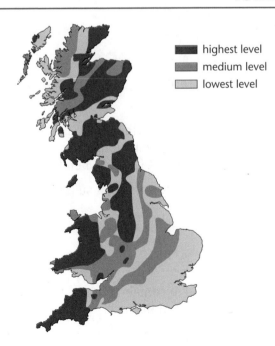

- ■ highest level
- ■ medium level
- ■ lowest level

a) Name **two** sources of background radiation.

   ....

   .... ②

Radioactivity

b) How can the level of background radiation be measured?

....

....

.... ④

c) Suggest why airline pilots and cabin crew are subjected to higher levels of background radiation than people who work on the ground.

....

.... ②

d) In the South-west of England the level of background radiation is high. This is due to the decay of radium-224 in the granite rock that lies underground.
Radium-224 decays to radon-220 with a half-life of 3.6 days.
Radon-220 decays by alpha emission with a half-life of 52 s.

i) Suggest why there is a plentiful supply of radium-224 in rocks that were formed millions of years ago.

....

.... ②

ii) Explain why the presence of radon-220 gas in buildings is a health hazard to people.

....

....

.... ③

TOTAL 13

❸ The table shows some radioactive materials found in rocks, with their half-lives and the substances formed when they decay.

Isotope	Substance formed on decay	Half-life in millions of years
rubidium-87	strontium-87	49 000
thorium-232	lead-208	14 000
uranium-238	lead-206	4500
potassium-40	argon-40	1250
uranium-235	lead-207	704
iodine-129	xenon-129	17

The age of the Earth is thought to be 4600 million years.
Rocks are dated by comparing the amount of an isotope present in a rock to the amount of the substance that is formed when it decays.

a) Explain which isotope is most suitable for dating rocks thought to have the same age as the Earth.

....

....

.... ③

b) A rock is found to contain thorium-232 and lead-208 in the ratio 1 : 3.
Estimate the age of the rock.

....

....

.... ③

c) Explain why rocks cannot be dated by comparing the proportions of iodine-129 and xenon-129.

....

....

.... ③

TOTAL 9

④ The graph represents the decay of a sample of americium.

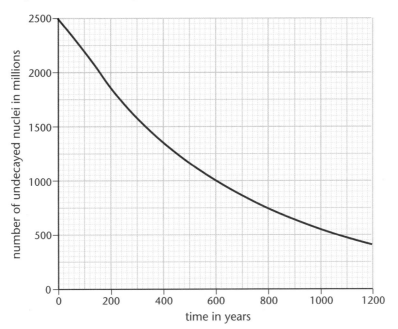

a) Determine the half-life of americium.
Show clearly how you do this on the graph.

....

.... ③

Americium is a radioactive isotope that emits alpha particles.
It is used in smoke alarms.

b) Describe the structure of an alpha particle.

.... ②

c) The diagram shows the structure of a smoke alarm.

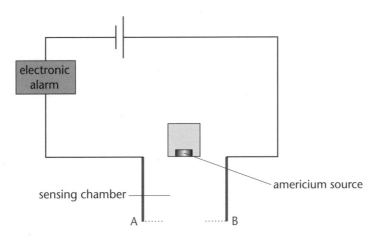

The alpha particles ionise the air between the electrodes A and B, allowing a current to pass in the sensing chamber.
If smoke enters the chamber, this current is stopped and the alarm sounds.

i) Explain how the current is stopped when smoke enters the chamber.

....

.... ②

ii) Explain why an alpha-emitting source is preferred to one that emits beta or gamma radiation for this purpose.

....

.... ②

iii) Why would an alpha-emitting source that has a half-life of 2 years not be suitable?

....

.... ②

TOTAL 11

QUESTION BANK ANSWERS

1 a) i) B ❶
   ii) C ❶
   iii) A ❶
b) The half-life is the average time ❶
   Taken for half the undecayed nuclei in a
   sample to decay ❶
c) i) Geiger–Müller tube and counter ❶
   ii) 260 counts/s ❶
     130 counts/s ❶

Examiner's tip

A common error is to state that all the material has decayed after two half-lives. This is not the case; after two half-lives the activity has dropped to $0.5 \times 0.5 = 0.25$ of the original activity.

   iii) Radioactive decay is random ❶
     Its activity at anytime could be higher or
     lower than the predicted value. ❶

2 a) Cosmic radiation (space)
The ground, buildings
Food/plants
Medical use of radioactive isotopes
Nuclear power stations.
       any two, 1 mark each ❷
b) Place a detector and counter away from
any radioactive materials ❶
Measure the number of counts over a
period of time, at least one minute.
Divide by the number of seconds. ❶
Repeat and take an average ❶
  Quality of written communication mark –
answer is structured in terms of sentences. ❶

Examiner's tip

Background radiation is sporadic and can vary significantly from second to second and from minute to minute. This is why it is important to take more than one reading and work out an average value.

c) Some cosmic radiation is absorbed by the
atmosphere ❶
So people at ground level are not
exposed to the level of cosmic radiation
that is experienced by people in aircraft ❶
d) i) Although radium-224 has a short
half-life ❶
It is probably formed from the decay
of an isotope with a longer half-life ❶

   ii) Radon-220 is a gas so it can enter the
lungs ❶
The alpha particles cannot penetrate
body tissue, but will be absorbed by lung
tissue ❶
They can damage cells and cause cancer ❶

Examiner's tip

Outside the body, alpha radiation is not very threatening to humans as it cannot penetrate the skin. But when it is inside the body, it can cause a lot of damage due to its intense ionising ability.

3 a) Uranium-238 ❶
This has a half-life similar to the age
of the Earth ❶
So it is possible to make a precise comparison
between the amount of uranium-238 and
lead-206 as they should be present in
approximately equal amounts. ❶
b) Three-quarters of the thorium has
decayed into lead ❶
So it has passed through two half-lives ❶
The age of the rock is $2 \times 14\,000$ million
years = 28 000 million years. ❶
c) Xenon is a gas ❶
And will seep out of the rock ❶
Therefore it is not possible to make a
reliable measurement of the amount of
xenon formed from the decay of iodine. ❶

4 a) Two readings taken of the time for the
number of undecayed nuclei to halve ❶
These readings averaged ❶
Answer in the range 440–460 years ❶

Examiner's tip

When taking readings from a graph, always draw lines with a ruler to show how you have done this.

b) An alpha particle consists of two neutrons ❶
And two protons ❶
c) i) The radiation emitted by the source
cannot penetrate the smoke ❶
So no ionisation of the air particles
occurs ❶
   ii) Alpha particles are easily absorbed and
would not penetrate the smoke ❶
Beta particles and gamma radiation
would pass through the smoke ❶
   iii) The activity would decrease noticeably
over the lifetime of the alarm ❶
It would need to be replaced frequently
to prevent a false alarm ❶

Radioactivity

Electronics

To revise this topic more thoroughly, see Chapter 7 in *Letts Revise GCSE Physics Study Guide.*

Try this sample GCSE question and then compare your answers with the Grade C and Grade A model answers on the next page.

Some householders fit a device that automatically switches some lights on at night and switches them off at dawn. This device can be used to deter burglars when the householders are on holiday.
The diagram shows a suitable circuit.

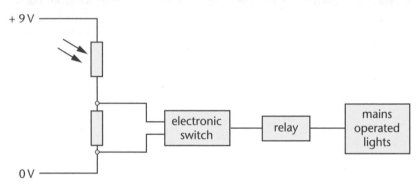

a   Explain how the voltage across the fixed resistor changes as the light level decreases.

....

.... **[3]**

b   The electronic switch is 'on' when the voltage across it is between 0 V and 3.0 V.
The resistance of the fixed value resistor is 1200 Ω.
Calculate the resistance of the light-dependent resistor when the lights switch on or off.

....

.... **[3]**

c   Describe and explain the effect of increasing the resistance of the fixed value resistor.

....

.... **[3]**

d   Explain why a relay is needed between the switching circuit and the mains operated lamps.

....

.... **[2]**

(Total 11 marks)

These two answers are at Grade C and A. Compare which one your answer is closest to and think how you could have improved it.

GRADE C ANSWER

One mark out of three is awarded for his first sentence, which is correct. The voltage across both resistors cannot go down. The resistors are in series, so the voltages must always add to 9 V.

Frank

a   As it gets darker the resistance of the LDR increases. ✓ The voltage across both resistors goes down.

b   The voltage across the LDR must be 6 V because voltages add across components in series. I cannot calculate the resistance because I do not know the current. ✓

Again, Frank shows that he does not understand how the voltage is shared between resistors in a series circuit.

c   Even less current passes in the switching circuit.

d   It enables a small direct voltage to switch a much higher voltage that can be either alternating or direct. ✓✓

One mark here, again for the first sentence. Frank should be able to calculate the current or work out that the voltages are in the same ratio as the resistances.

An excellent answer – full marks.

4 marks = Grade C answer

Grade booster ⋯⋯▷ move a C to a B
Frank has shown that he has some knowledge of the voltages across resistors in series, but his understanding of how this operates is poor. Had he been able to apply his knowledge, he could have gained two or three more marks to move his grade to a B.

GRADE A ANSWER

Francesca gains full marks here. She has explained the effect on the voltages correctly.

Francesca

a   The resistance of the LDR goes down, so its share of the voltage also goes down. This means that the voltage across the fixed value resistor must go up. ✓✓✓

b   The voltage across the LDR is 6 V. As this is twice the voltage across the fixed value resistor, it must have twice the resistance. So the answer is 2400 Ω. ✓✓✓

Only two marks out of three. The reasoning in the first sentence is correct. However, for the third mark she needs to state whether the switch would operate at a lower or a higher light intensity.

c   It has a bigger share of the resistance, so it has a bigger share of the voltage. This means that the switch will operate at a different light intensity. ✓✓

d   It relays the information from one circuit to the next. ✗

Full marks again. Francesca shows a high level of understanding of how the voltages in a series circuit are related to the values of the resistors.

This answer is too vague to be awarded any marks.

8 marks = Grade A answer

Grade booster ⋯⋯▷ move A to A*
Francesca's understanding of this topic is certainly at A* level, but she lets herself down with her lack of knowledge about the purpose of a relay. A high level of knowledge, understanding and data handling skills are all needed to gain an A* grade.

Electronics

1 A burglar alarm is operated by three switches. Switches A and B are fitted to two separate doors and switch M is the master switch. When a door is opened or the master switch is turned on it causes the input to a logic gate to be 'on'. The diagram shows the circuit that is used.

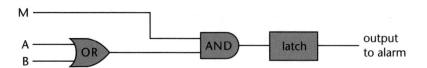

a) Complete the table using a 0 for 'off' and a 1 for 'on'.

	Inputs to OR gate		Inputs to AND gate		
	A	B	OR gate output	M	AND gate output
i)	0	0		1	
ii)	1	0		1	
iii)	0	1		0	
iv)	1	1		0	

④

b) Which line of the table shows what happens when the master switch is 'on' and the doors are closed?

.... ①

c) Which lines of the table show what happens when a door is opened while the master switch is 'off'?

.... ①

d) Describe what has to happen for the alarm to sound.

....
.... ②

e) Explain the purpose of the latch.

....
.... ②

TOTAL 10

2 The diagram represents an electronic system that is used to control the temperature in an incubator.

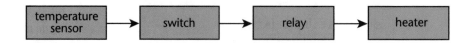

a) What is the purpose of the relay between the switch and the heater?

....
.... ②

The circuit diagram shows how the temperature sensor is made from a fixed resistor and a thermistor. The resistance of the thermistor decreases when its temperature increases.

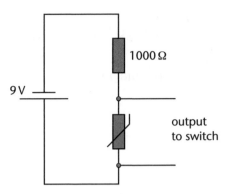

b) The switch is switched on when the input to it is 0.6 V or more.

  i) Calculate the voltage across the 1000 Ω resistor when the voltage across the thermistor is 0.6 V.

  .... ①

  ii) Use your value from i) to calculate the current in the circuit and the resistance of the thermistor when the voltage across the thermistor is 0.6 V.

  ....

  ....

  ....

  .... ④

  iii) Suggest why the current in the temperature sensing circuit needs to be low.

  ....

  .... ②

c) Explain how the switch turns the heater off when the temperature of the thermistor rises.

  ....

  ....

  .... ③

TOTAL 12

3 The diagram shows a network of five resistors.

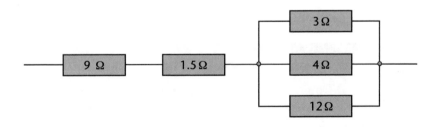

a) Calculate the total resistance of the three resistors connected in parallel.

  ....

  .... ④

b) Calculate the total resistance of the two resistors connected in series.

  ....

  .... ②

c) Calculate the total resistance of the network.

  .... ①

Electronics

d) A 9 V battery is connected in series with the network.
Complete the table to show the current in each resistor and the voltage across each resistor.

Resistor value in Ω	Current in A	Voltage in V
9		
1.5		
3		
4		
12		

⑤

TOTAL 12

④ Logic gates can be used to switch electrical devices on and off according to environmental conditions. A lamp fitted to the outside of a house is controlled by logic gates. The inputs to the gates come from a light sensor and an infrared sensor. The tables show the inputs to the logic gates from the sensors.

Light condition	Input from light sensor
light	1
dark	0

Infrared condition	Input from infrared sensor
high	1
low	0

a) What could cause the amount of infrared radiation to change from 'low' to 'high'?

.... ①

b) The diagram shows the circuit that is used.

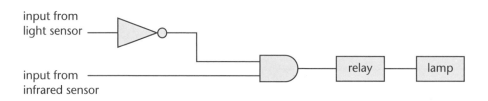

i) Complete the table to show the state of the output from the AND gate.

Light condition	Infrared condition	AND gate output
dark	low	
light	high	
dark	high	
light	low	

④

ii) Describe the conditions necessary for the lamp to be switched on.

....

.... ②

c) A householder wishes to use this circuit to switch on an outside lamp while he puts his car in the garage after returning home at night.

  i) Explain why the circuit shown in the diagram is not suitable for this purpose.

  ....

  .... ①

  ii) Describe and explain how a bistable circuit could be added to make the circuit suitable.

  ....

  ....

  .... ③

  TOTAL 11

⑤ The diagram shows a time delay circuit.

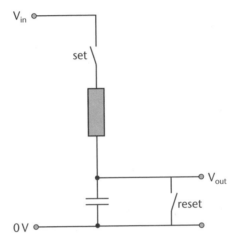

a) Describe and explain how the output voltage changes when the 'set' switch is closed.

  ....

  ....

  .... ③

b) State **two** ways in which the time delay can be increased.

  ....

  .... ②

c) Explain why the 'reset' switch is necessary.

  ....

  .... ②

  TOTAL 7

1 a) i) OR 0   AND 0   ❶
    ii) OR 1   AND 1   ❶
    iii) OR 1   AND 0   ❶
    iv) OR 1   AND 0   ❶

Examiner's tip

Unless otherwise stated, an OR gate has a '1' output when either or both of the inputs are '1'.

b) Line i)   ❶
c) Lines iii) and iv)   ❶
d) Either door A or door B has to be opened   ❶
    And the master switch has to be on   ❶

Examiner's tip

The condition that either door A or door B has to be opened includes the possibility of both being opened at once.

e) The latch keeps the alarm switched on   ❶
    After a door has been closed   ❶

2 a) The current in the heater is too great for the switch to handle.   ❶
    The relay enables the large current in the heater to be switched by a much smaller current.   ❶
b) i) 9 V – 0.6 V = 8.4 V   ❶

Examiner's tip

Note that in a series circuit the total voltage is the sum of the voltages across the components.

    ii) $I = V \div R$ ❶
      Circuit current = 8.4 V ÷ 1000 Ω   ❶
      = 0.0084 A or 8.4 mA (8.4×10^{-3} A)   ❶
      Thermistor resistance = 0.6 V ÷ 0.0084 A
      = 71.4 Ω   ❶

Examiner's tip

This assumes that the amount of current that passes into the switch is so small it can be neglected, so that the current in the fixed resistor and the thermistor is the same.

iii) The current in the thermistor has a heating effect   ❶
     This needs to be minimized so that it does not turn the heater off when the temperature is too low   ❶
c)   The 9 V supply voltage in the sensing circuit is shared between the thermistor and the fixed resistor.   ❶
    When the thermistor gets warmer its resistance drops and its share of the voltage goes down.   ❶
    The switch is turned off when the voltage across the thermistor drops below 0.6 V   ❶

3 a) The resistance of the three resistors in parallel is calculated using
$$\frac{1}{R} = \frac{1}{R_1} + \frac{1}{R_2} + \frac{1}{R_3}$$ ❶
$$= \frac{1}{3\Omega} + \frac{1}{4\Omega} + \frac{1}{12\Omega} = \frac{8}{12\Omega}$$ ❶
$$R = \frac{12\Omega}{8} = 1.5\,\Omega$$ ❶

Quality of written communication mark – answer set out in a clear, logical format.   ❶

Examiner's tip

When calculating the total resistance of a number of resistors in parallel, remember to reciprocate the calculated value of 1/R. The common error is to omit the last step, giving an answer here of 0.67 Ω.

b) 9 Ω + 1.5 Ω   ❶
    = 10.5 Ω   ❶

Examiner's tip

The first mark here is for appreciating that when resistors are connected in series, the total resistance is equal to the sum of the individual resistor values.

c) 10.5 Ω + 1.5 Ω = 12.0 Ω   ❶
d) The correctly completed table is:

Resistor value in Ω	Current in A	Voltage in V
9	0.75	6.75
1.5	0.75	1.125
3	0.38	1.125
4	0.28	1.125
12	0.09	1.125

1 mark for each correct row   ❺

❹ a) Any warm object such as a person or a car
   engine   ❶

b) i) 0   ❶
      0   ❶
      1   ❶
      0   ❶

Examiner's tip

▶▶▶ *Both inputs have to be '1' for the output of an
AND gate to be '1'.*

   ii) It must be dark   ❶
       And there must be a warm object to cause
       the infrared to be 'high'   ❶
c) i) The lamp goes off when he drives the car
      into the garage   ❶
   ii) A bistable latch could be placed between
       the AND gate and the relay   ❶
       This would keep the lamp switched on ❶
       After the output from the AND gate had
       changed to 0   ❶

Examiner's tip

▶▶▶ *You should also remember that a bistable latch
needs to have a means of resetting it.*

❺ a) Current passes in the resistor   ❶
     This causes the capacitor to charge   ❶
     As it charges the voltage across it rises ❶
   b) Increase the resistance of the resistor ❶
      Increase the capacitance of the capacitor ❶

Examiner's tip

▶▶▶ *The rate at which the capacitor charges is
determined by both its capacitance and the
resistance of the charging circuit.*

   c) The reset switch discharges the capacitor ❶
      This is necessary so that the circuit can be
      used again   ❶

Electronics

CHAPTER 8

Using waves

To revise this topic more thoroughly, see Chapter 8 in *Letts Revise GCSE Physics Study Guide*.

 Try this sample GCSE question and then compare your answers with the Grade C and Grade A model answers on the next page.

Radio and television aerials are designed to absorb energy from waves of a particular wavelength and frequency. VHF radio aerials are designed to receive waves with a frequency of approximately 1.0×10^8 Hz.

a The speed of radio waves in air is 3.0×10^8 m/s. Calculate the wavelength of a radio wave that has a frequency of 1.0×10^8 Hz.

....

....

.... **[3]**

b Radar uses short wavelength radio waves emitted in a narrow beam. The diagram shows how this is achieved by reflecting the beam from a reflector onto a transmitting dish.

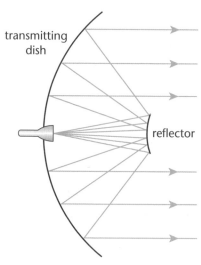

The larger the diameter of the dish that is used, the less the beam spreads out.

(i) Suggest why narrow beams of radiation are used for radar.

....

.... **[1]**

(ii) Explain why a large diameter dish has to be used to reduce the beam spread.

....

.... **[2]**

(Total 6 marks)

These two answers are at Grade C and A. Compare which one your answer is closest to and think how you could have improved it.

GRADE C ANSWER

Camilla

Camilla is awarded two marks out of three for this response. She has remembered the relationship, transposed it correctly and made a correct substitution of the values. She has then worked out the answer correctly, but not given a unit for the wavelength.

a I have worked out that wavelength = speed ÷ frequency ✓ so the wavelength is $3.0 \times 10^8 \div 1.0 \times 10^8$ ✓ = 3

b (i) So that the energy of the beam has not spread out too much when it reaches its target, such as a ship or an aircraft. ✓

(ii) This is because a small diameter dish would not reduce the beam spread as much as a large diameter dish does. It's all to do with refraction. ✗

Camilla gains the mark here, the energy of a radar beam needs to be concentrated.

Camilla has only repeated the information given in the question, in a slightly different form. She has then confused 'refraction' with 'diffraction'. No marks.

3 marks = Grade C answer

Grade booster ····▷ move a C to a B
Grade C candidates rarely show any understanding of topics such as diffraction. To improve this, a grade C candidate should read as much as possible about the topic and practise answering questions.

GRADE A ANSWER

Charles

Full marks for this answer. Notice how Charles sets his answer out in a clear, logical way and gives the correct units for all the physical quantities.

a $\lambda = v \div f = 3.0 \times 10^8$ m/s $\div 1.0 \times 10^8$ Hz ✓✓ = 3.0 m ✓

b (i) The narrower the beam, the more concentrated it is. As it does not spread out as much as a wide beam, a lower power is needed. ✓

Full marks again. Charles has actually given a fuller answer than the question demands when he states that 'a lower power is needed'.

One mark out of two. Charles shows that he can apply his knowledge of diffraction to what may be an unfamiliar situation. Only the highest-achieving candidates are able to do this.

(ii) Waves spread out when they leave an opening such as a transmitting dish. This is diffraction. To reduce the spreading out, the diameter of the dish needs to be bigger than the wavelength of the waves that are being transmitted. ✓

5 marks = Grade A answer

Grade booster ····▷ move A to A*
Charles' answer is almost faultless. The only mark that he failed to gain was for stating that the dish diameter needs to be many times the wave length in **b (ii)**. This is what examiners are looking for to distinguish between grade A candidates and grade A*.

Using waves

1

a) National television and radio broadcasts are transmitted from the Telecom tower, a very tall building in London. The sound and picture information is carried by microwaves that cannot be detected by a domestic radio or television set. One reason for using microwaves is that they are more easily focused into a narrow beam.

   i) Calculate the wavelength of microwaves which have a frequency of 1.1×10^{10} Hz. The speed of microwaves in air is 3.0×10^8 m/s.

   ....

   ....

   .... ③

   ii) Describe **one** similarity and **one** difference between microwaves and the radio waves that are detected by a domestic radio.

   ....

   .... ③

b) Repeater stations receive the microwaves and send them on to other repeater stations. Repeater stations are necessary because the microwaves cannot be received more than 50 km from their source. The diagram shows part of the network of repeater stations around the south of England.

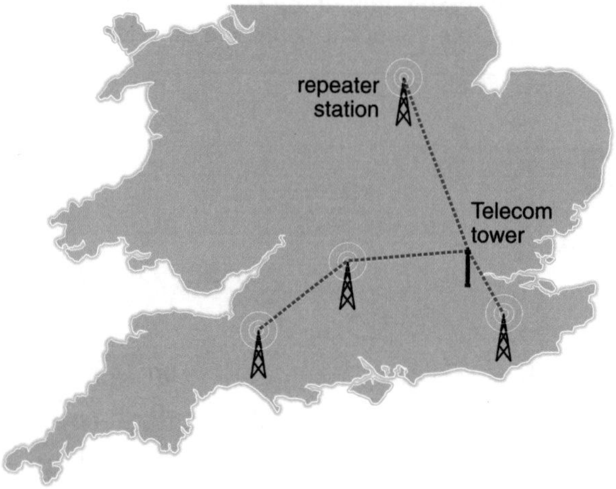

   i) Explain why the microwaves cannot travel long distances over the Earth's surface.

   ....

   .... ②

   ii) Suggest why the microwaves are transmitted as a narrow beam.

   ....

   .... ②

   TOTAL 10

ANSWERS ON PAGE 82 ANSWERS ON PAGE 82 ANSWERS ON PAGE 82 ANSWERS ON PAGE 82

2 Microwave cookers use radio waves of wavelength 0.12 m to heat food. The speed of the waves is 3.0×10^8 m/s.

a) Calculate the frequency of the waves used in microwave cooking.

....

....

.... ③

b) The microwaves cause water molecules in food to resonate.

　i) Describe what happens when the water molecules resonate.

....

....

.... ③

　ii) Explain how this causes the food to become hot.

....

.... ②

　iii) Pyrex dishes and other food containers are not heated by the microwaves. What does this tell you about the frequency at which the particles in Pyrex oscillate?

.... ①

....

TOTAL 9

---

3 The diagram shows two radio transmitters. The transmitters are broadcasting identical radio waves of wavelength 3.0 m.

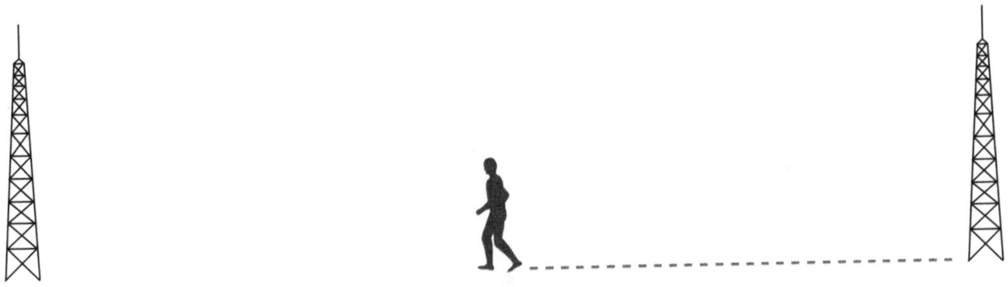

A person walks between the transmitters, listening to the broadcast on a walkman. He notices that the sound gets louder, then quieter, then louder and so on as he walks. When he stands still the loudness of the sound stays the same, but there are some places where the sound is loud and other places where it is quiet.

a) Use diagrams to explain how the radio waves from the transmitters combine to give a very strong radio signal at some points and a weak signal at others.

④

b) Suggest why this could be a nuisance to a householder who lives within the range of both transmitters.

....

.... ②

TOTAL 6

4 The diagram represents an object placed at a distance of 5.0 cm from a convex lens of focal length 8.0 cm.

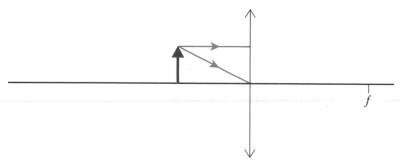

a) Complete the ray diagram and draw in the image. ③

b) What is the distance between the lens and the image?

.... ①

c) State **two** differences between the object and the image.

....

.... ②

d) Which optical instrument uses a convex lens to produce this type of image?

.... ①

e) Describe what happens to the size and position of the image when the object is moved towards the lens.

....

.... ②

TOTAL 9

5

a) The diagram shows a satellite in a low polar orbit.

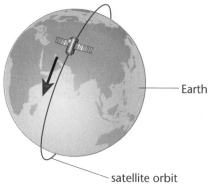

Earth

satellite orbit

i) The satellite completes one orbit in 90 minutes.

Explain how the satellite is able to view the whole of the Earth's surface in one day.

....

....

.... ③

ii) Explain why this is a useful orbit for weather satellites but not for communications satellites.

....

....

.... ③

b) Some communications satellites occupy geostationary orbits.
State **three** differences between a low polar orbit and a geostationary orbit.

....

....

.... ③

c) Communication with satellites uses very short wavelength radio waves.
Explain why longer wavelength waves are not used.

....

.... ②

TOTAL 11

6 A transducer transfers a signal in one form to a signal in another form.

a) Write down the name of a transducer that transfers:

    i) a signal from a sound wave into an electrical signal

.... ①

    ii) an electrical signal into sound

.... ①

    iii) an electrical signal into a magnetic signal

.... ①

    iv) a magnetic signal into an electrical signal.

.... ①

b) The diagram shows how the information stored on a compact disc (CD) is read.

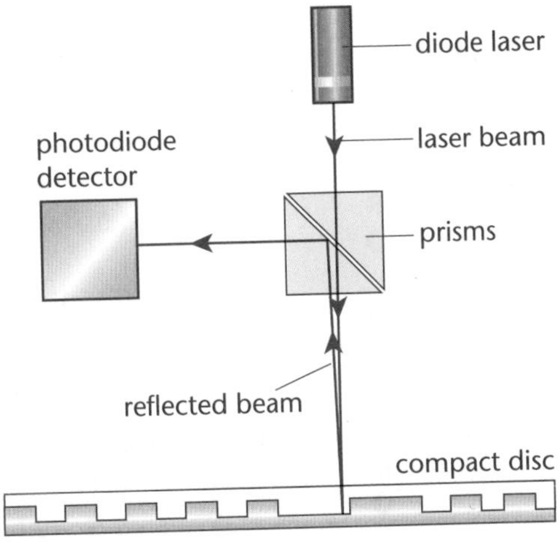

    i) Describe how information is stored on a compact disc.

....

....

.... ③

    ii) Explain how the diode laser and photodiode detector read this information.

....

....

.... ③

TOTAL 10

QUESTION BANK ANSWERS

1 a) i) Wavelength = speed ÷ frequency ❶
= 3×10^8 m/s ÷ 1.1×10^{10} Hz ❶
= 0.027 m ❶

Examiner's tip
Note that the first mark is for knowledge and correct transposition of the wave equation, $v = f \times \lambda$.

ii) Similarity: either transverse OR electromagnetic OR speed ❶
Difference: either wavelength OR frequency ❶
Quality of written communication mark – correct spelling of scientific terms. ❶

Examiner's tip
All electromagnetic waves are transverse and travel at the same speed in a vacuum. They differ in wavelength and frequency.

b) i) Microwaves travel in straight lines ❶
The Earth's surface is curved ❶
ii) So that the maximum power is received at the next dish ❶
Narrow beams enable small dishes to be used ❶

2 a) Frequency = speed ÷ wavelength ❶
= 3.0×10^8 m/s ÷ 0.12 m ❶
= 2.5×10^9 Hz ❶
b) i) Water molecules in the food are vibrating ❶
The frequency of the microwaves is the same as the frequency of vibration of the water molecules ❶
This causes increased amplitude of vibration of the water molecules ❶

Examiner's tip
This is an example of resonance being used to do a useful job.

ii) The water molecules have increased energy ❶
This energy is passed on by conduction to the other molecules in the food ❶

Examiner's tip
Temperature or 'hotness' is a measure of the energy that the particles of a substance have.

iii) It is not the same as the frequency of the microwaves ❶

3 a) When two wave crests or two wave troughs meet they combine to give a wave of increased amplitude ❶

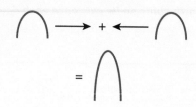

❶

Where a wave crest meets a wave trough they combine to give a wave of reduced amplitude ❶

❶

b) The householder could have a very weak signal ❶
This would happen if the house were at a place where the waves interfered destructively ❶

4 a) The ray to the centre of the lens continued straight ❶
The ray parallel to the principal axis drawn so that it passes through the principal focus ❶

These rays traced back and an upright arrow drawn where they cross. ❶
b) Answer in the range 12–14 cm. ❶

Examiner's tip
The correct answer is 13.3 cm, but this level of precision is not expected from a ray diagram.

c) The object is real and the image is virtual ❶
The image is larger than the object ❶
d) A magnifying glass ❶

Examiner's tip
Most optical instruments produce a real image that is on the opposite side of the lens to the object. A magnifying glass produces a virtual image on the same side of the lens as the object.

e) The image becomes closer to the lens ❶
And smaller ❶

5 a) i) The satellite completes 16 orbits in one
day   ❶
While it is completing an orbit, the Earth
is turning round   ❶
So the satellite sees a different view on
each of its orbits   ❶

Examiner's tip

*The Earth spins on its axis once each day, so for
each orbit of the satellite it turns through 360°
÷ 16 = 22.5°.*

  ii) Weather satellites need to see different
views so that they can see the weather that
is approaching/monitor changing weather
conditions   ❶
Communications satellites need to be in a
fixed point relative to the Earth   ❶
So that the transmitting and receiving
dishes do not need to track them/they can
be used for communications between two
places 24 hours a day   ❶
b) A geostationary orbit:
is higher than a low polar orbit   ❶
is above the equator   ❶
remains at a fixed point above the Earth   ❶

Examiner's tip

*Satellites in geostationary orbits have an orbit
time of 24 hours, the same time as the length of
a day. This means that they complete one orbit
in the time it takes the Earth to spin once on its
axis. The satellites remain above the same point
on the Earth's surface. However, it is only
possible for a satellite to be geostationary if it
is in orbit at a certain height above the equator.*

  c) Short wavelength radio waves can pass
through the ionosphere as space waves.   ❶
Longer wavelength radio waves would be
reflected by the ionosphere.   ❶

6 a) i) A microphone   ❶

Examiner's tip

*A microphone produces an electrical copy of a
sound signal. Some microphones do this by
transferring some of the energy from the sound
wave into electricity. Most modern microphones
have an internal battery that acts as the energy
source for the electric current.*

  ii) A loudspeaker   ❶

Examiner's tip

*Remember, sound is kinetic energy of the air
particles that transmit it.*

  iii) A tape record head   ❶
  iv) A tape playback head   ❶
b) i) In digital form   ❶
As a series of 'pits'   ❶
And 'bumps'   ❶

Examiner's tip

*Information is stored either in analogue form or
digitally. Vinyl records can only store
information in an analogue form, and compact
discs can only store information digitally.
Magnetic tape can be used to store information
in either form.*

  ii) Radiation from the diode laser passes through
the prisms onto the CD   ❶
It is reflected back to the photodiode
reflector   ❶
The detector recognises from the time delay
whether it is reading a 'pit' or a 'bump'   ❶

Examiner's tip

*Most of the information that you need to
answer this question is given in the stem. Make
sure that you take time to read the stem of a
question thoroughly.*

CHAPTER 9

Forces and their effects

To revise this topic more thoroughly, see Chapter 9 in *Letts Revise GCSE Physics Study Guide.*

> Try this sample GCSE question and then compare your answers with the Grade C and Grade A model answers on pages 85 and 86.

Car manufacturers test the safety of their vehicles by driving them into brick walls. During such an experiment, the vehicle carries a 'dummy' driver and passengers.
The diagrams show the effects on the 'dummies' during such a collision.

The total mass of the car and its occupants is 875 kg.
They are brought to rest from a speed of 15 m/s in a time interval of 0.5 s.

a Calculate the size of the force needed to stop the car.

....

....

.... **[3]**

b What force would be needed to bring the 65 kg driver to a halt in the same time interval?

....

.... **[2]**

c Explain how the stretching of the seat belt reduces the force needed to stop the driver.

....

.... **[2]**

d The child in the back of the car appears to have been pushed forwards.

Television advertisements claim that in these circumstances she is 'pushed forwards with the force of a baby elephant'.

Criticise this statement.

....

....

.... **[3]**

e Modern cars have 'crumple zones' and are fitted with air bags.

Explain how these help to protect the occupants during a collision.

....

....

....

.... **[4]**

(Total 14 marks)

These two answers are at Grade C and A. Compare which one your answer is closest to and think how you could have improved it.

GRADE C ANSWER

Bronwen is awarded two marks out of three. She has recalled the relationship and made a correct substitution of the values. She also has the correct numerical answer, but she has given the wrong unit, so she loses the final mark.

No marks here. Bronwen has not appreciated that the deceleration is less, so less force is needed.

One mark out of four for stating that 'the pressure is less'. This is because the force on the driver acts over a greater area.

Bronwen

a Force = mass × acceleration ✓
= 875 × (15 ÷ 0.5) ✓ = 26 250 kg.

b It would be less because the mass is less. ✗
The force would only be 65 × (15 ÷ 0.5)
= 1950 kg. ✓✓

c When the seat belt stretches it takes the force from the driver, so less is needed. ✗

d Well this is right because it's as if something has hit you hard in your back. It could be a baby elephant, you wouldn't know the difference. ✗

e The crumple zones crumple up. Air bags fill with gas and surround the driver, so the pressure on the driver is less. ✓ They act like cushions.

Two marks out of two here. The calculation is correct but again she has the wrong unit. However, she is not penalised twice for the same error, so she gains full marks on this part of the question.

Bronwen is now out of her depth. She clearly has some knowledge or experience about this situation, but she cannot relate it to her knowledge and understanding of physics.

5 marks = Grade C answer

Grade booster ····⟩ move a C to a B
Bronwen has shown that she can calculate the size of a force using $F = m \times a$. However, to gain a grade B she would be expected to know that force is measured in N and to show some understanding of this relationship by applying her knowledge to explain the action of a seat belt.

Full marks are awarded here for a correct calculation and unit.

Brian

a Force = mass × acceleration ✓
= 875 kg × (15 m/s ÷ 0.5 s) ✓
= 26 250 N. ✓

*Full marks again. Notice how Brian has worked this out using ratios rather than repeating the calculation in **a**. This shows a high level of understanding.*

b As the mass is only 65 kg, the force is 65/875 of the answer to a = 1950 N. ✓✓

Full marks again. Brian has not quite put all the steps into the argument – he could have stated that the driver takes a longer time to stop – but he clearly understands the principle involved.

c If the seat belt stretches, the driver carries on moving after the car has stopped. ✓ The deceleration is less, so the force is less. ✓

Two marks are awarded for the statements in the first sentence. Brian has explained why the child carries on moving. He is not able to make a criticism about the baby elephant. The physics of this statement is nonsense, because there is no such force and there is no such thing as 'the force of a baby elephant'.

d I don't think she has been pushed forwards, she just carries on moving because she doesn't have a seat belt or anything to stop her. ✓✓ Also, it doesn't say what the force of a baby elephant is. And there can be different sizes of elephant babies.

e Crumple zones act like seat belts. They prolong the collision time and so make the force on the driver smaller. ✓✓ Air bags probably do something similar.

Two marks for the statement about crumple zones. Brian does not understand the action of an air bag – it is quite different from a crumple zone – so he is not awarded any marks for this response.

11 marks = Grade A answer

Grade booster ⋯⟩ move A to A*

An A* candidate is expected to be apply his/her understanding of physical principles to situations that may be familiar or previously unknown. Brian has shown that he has the ability to use ratios where this is appropriate, but his application of his knowledge is not good enough for him to be awarded A*.

1 A light plastic football that has a mass of 0.10 kg is thrown at a 0.6 kg 'skittle'.

8.0 m/s

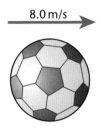

a) The football travels towards the skittle at a speed of 8.0 m/s. Calculate the momentum of the ball.

....

....

.... ②

b) After being hit by the ball, the skittle moves to the right with a speed of 2.0 m/s. Calculate the momentum of the skittle and the momentum of the football after the ball hits the skittle.

....

....

....

.... ④

c) Calculate the speed of the football after it hits the skittle, and state the direction of travel.

....

....

....

.... ③

d) Explain why the bowls used in ten-pin bowling are much more massive than the 0.1 kg ball.

....

....

.... ③

TOTAL 12

2 In an archery contest, the contestants fire arrows at a round target.
They aim at a target that is 70 m away from the contestants.
The 'bulls-eye' is 1.3 m above the ground.
Assume that $g = 10$ m/s².

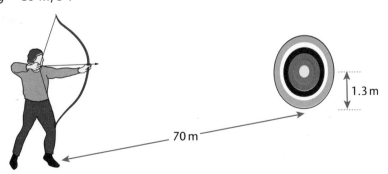

70 m

1.3 m

An archer fires an arrow horizontally at a speed of 190 m/s.

Forces and their effects

a) Describe and explain the path of the arrow as it travels through the air.

....

....

.... ③

b) Calculate the time it takes for the arrow to reach the target.

....

....

.... ③

c) Calculate the vertical distance that the arrow falls through as it travels through the air.

....

....

.... ④

d) Explain whether the arrow hits the target.

....

.... ②

e) What should the archer do to hit the 'bulls-eye'?

.... ①

TOTAL 13

---

3 The diagram shows a satellite in a low polar orbit.

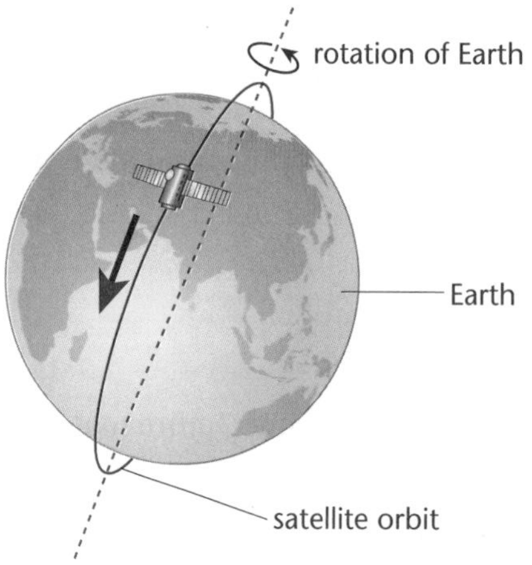

a) Why is this orbit described as 'polar'?

.... ①

b) The satellite completes one orbit of the Earth in 90 minutes.
How many orbits does it complete in one day?

.... ①

c) A satellite in a low polar orbit is used to monitor changing weather patterns.
Explain why this is a suitable orbit for a weather satellite.

....

....

.... ③

d) State **one** other use for a satellite in a low polar orbit.

.... ①

ANSWERS ON PAGE 90 ANSWERS ON PAGE 90 ANSWERS ON PAGE 90 ANSWERS ON PAGE 90

e) Explain why a low polar orbit is not suitable for a satellite that broadcasts television programmes from America to Europe.

....

....

.... ③

f) The radius of the satellite's orbit is 6.7×10^6 m.

i) Calculate the speed of the satellite in its orbit.

....

.... ②

ii) Calculate the satellite's acceleration.

....

.... ②

iii) What is the direction of the satellite's acceleration?

.... ①

iv) On the Earth's surface the value of free-fall acceleration, g, is 9.8 m/s².

What is the value of g at the height of the satellite's orbit?

.... ①

TOTAL 15

④ The diagram shows a traffic cone and a bowling pin tilted to the same angle.
The X shows the position of the centre of mass.

a) Explain why the centre of mass of the cone is lower than that of the bowling pin.

.... ①

b) Describe and explain what happens to each object when it is released.

....

....

....

.... ④

c) Give **two** ways in which the design of the bowling pin could be changed to make it more difficult to knock over.

....

.... ②

TOTAL 7

Forces and their effects

❶ a) Momentum = mass × velocity
   = 0.1 kg × 8 m/s   **❶**
   = 0.8 Ns   **❶**

Examiner's tip

An alternative acceptable unit for momentum is kg m/s.

b) Momentum of skittle = 0.6 kg × 2.0 m/s   **❶**
   = 1.2 Ns   **❶**
   The total momentum of the skittle and ball is 0.8 Ns   **❶**
   Therefore the ball has momentum – 0.4 Ns   **❶**

Examiner's tip

Momentum is a vector quantity; it can have both positive and negative values.

c) Velocity = momentum ÷ mass   **❶**
   = – 0.4 Ns ÷ 0.10 kg   **❶**
   = – 4.0 m/s   **❶**

Examiner's tip

The first mark here is for transposing (rearranging) the relationship correctly. In the final answer, the negative sign shows that the football is moving from right to left.

d) The balls need to have a lot of momentum   **❶**
   This is so that they can knock several pins down   **❶**
   And still keep moving in the same direction   **❶**

❷ a) The path is curved, with the slope of the curve increasing   **❶**
   This is because the arrow travels equal distances in equal time intervals horizontally   **❶**
   But increasing distances in successive time intervals vertically   **❶**

Examiner's tip

This is the path of a projectile. All projectiles with a horizontal motion follow a parabolic path.

b) Time = distance ÷ speed   **❶**
   = 70 m ÷ 190 m/s   **❶**
   = 0.37 s.   **❶**
c) $s = ut + \frac{1}{2}at^2$ **❶**
   $= \frac{1}{2} \times 10 \text{ m/s}^2 \times (0.37s)^2$ **❶**
   = 0.68 m.   **❶**
   Quality of written communication mark – answer set out in a clear, logical form.   **❶**

Examiner's tip

The question is about the vertical distance travelled, so the initial velocity is 0. A common error at this level is to use the horizontal velocity as the initial velocity.

d) The arrow does hit the target   **❶**
   As the distance it falls is less than the height of the 'bulls-eye' above the ground   **❶**
e) The archer should aim at a distance of 0.68 m above the 'bulls-eye'   **❶**

❸ a) The orbit passes over the poles   **❶**
b) Sixteen   **❶**
c) While the satellite is completing each orbit, the Earth is rotating   **❶**
   So the satellite sees a different 'strip' of the Earth on its next orbit   **❶**
   In this way it views the whole of the Earth's surface in one day   **❶**

Examiner's tip

The Earth rotates through an angle of 22.5° during each orbit of the satellite.

d) For surveillance   **❶**
e) It would only be in a suitable position for part of its orbit   **❶**
   It would need to be tracked by the transmitting and receiving aerials   **❶**
   A satellite that stays at a fixed point above the Earth's surface is needed for television broadcasts   **❶**

Examiner's tip

Satellites used for television broadcasts occupy geostationary orbits. These have a time period of 24 hours, so they rotate at the same rate as the Earth. It is only possible to have a geostationary orbit in the plane of the equator.

f) i) Speed = 2π × radius ÷ time

          = $2\pi \times 6.7 \times 10^6$ m ÷ (90×60) s ❶

          = 7.8×10^3 m/s ❶

Examiner's tip

There is no mark for recall of the relationship as it is provided on the question paper.

ii) Acceleration = (speed)^2 ÷ radius

        = $(7.8 \times 10^3$ m/s$)^2$ ÷ 6.7×10^6 m ❶

        = 9.1 m/s^2 ❶

iii) Towards the centre of the Earth ❶

iv) 9.1 m/s^2 ❶

Examiner's tip

This question is testing whether you know that the centripetal acceleration of a satellite is its free-fall acceleration. A common error is to think that centripetal acceleration is in addition to free-fall acceleration.

❹ a) The cone is narrow at the top and wide at the bottom, so most of its mass is near the bottom of the cone. ❶

b) The cone returns onto its base ❶

Its weight has a clockwise turning effect or moment ❶

The bowling pin topples ❶

Its weight has an anticlockwise turning effect or moment ❶

Examiner's tip

The weight of an object acts from its centre of mass. If the arrow that represents the weight of an object passes through the base, the object is stable. If the arrow falls outside the base, the object will topple.

c) Make the base wider ❶

Make the bottom heavier, so that the centre of mass is lowered ❶

Examiner's tip

The effect of both these changes is that the bowling pin has to lean at a greater angle to make the weight arrow fall outside the base.

Forces and their effects

Particles

To revise this topic more thoroughly, see Chapter 10 in *Letts Revise GCSE Physics Study Guide.*

> Try this sample question and then compare your answer with the Grade C and Grade A model answers on the next page.

Atoms contain *protons*, *neutrons* and *electrons*. Each nucleon contains three quarks.

a   Which of the particles in *italics* are fundamental particles?

Explain why the others are not fundamental particles.

....

.... **[2]**

b   The diagram shows the structure of a proton and a neutron.

neutron   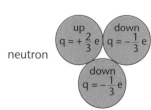   proton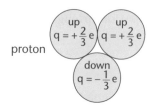

Describe what happens to a nucleus:

(i) in β^+ decay

....

.... **[3]**

(ii) in β^- decay.

....

.... **[3]**

c   Isotopes that are naturally radioactive do not decay by gamma emission only. Gamma emission usually follows decay by alpha or beta emission.

Explain the emission of gamma radiation.

....

.... **[2]**

(Total 10 marks)

These two answers are at Grade C and A. Compare which one your answer is closest to and think how you could have improved it.

GRADE C ANSWER

Charlie

Two marks out of two are awarded here for a correct answer.

a Electrons are fundamental particles but the others are made up of quarks so they cannot be fundamental. ✓✓

Only one mark here. Charlie knows that it is caused by the quark structure changing, but he is not able to use the diagram to work it out correctly.

b (i) A quark in a neutron changes from down to up, so it becomes a proton. ✓

One mark here, for having the opposite change to that in **b (i)**.

(ii) A quark in a proton changes from up to down, so it is now a neutron. ✓

c (i) This happens when a nucleus is left with too much energy, which it loses as a gamma ray. ✓

One mark for knowing that gamma radiation is emitted when a nucleus has excess energy. For the second mark he needs to make a statement about the nature of gamma radiation.

5 marks = Grade C answer

Grade booster ····⟩ move a C to a B
Charlie does not know the change in quark structure associated with beta decay, but had he studied the information in the diagram he could have worked it out from that. Instead of rushing in to answer the questions, he should take time to study the information that he has been given.

GRADE A ANSWER

Heather

Full marks for this correct answer.

a A fundamental particle is not made up of any other particles. Only the electron is a fundamental particle. ✓✓

b (i) An up quark in a proton changes to a down quark as it emits a positron. The number of protons goes down by one and the number of neutrons goes up by one. ✓✓✓

Heather has given a full description of the changes in the nucleus and is awarded three marks out of three.

Correct again. Three marks out of three.

(ii) A down quark in a neutron changes to an up quark, emitting an electron. The number of neutrons goes down by one and the number of protons goes up by one. ✓✓✓

c Gamma radiation comes from the nucleus. ✓

One mark is awarded here for the correct statement about gamma radiation. Heather has not stated why the nucleus emits gamma radiation.

9 marks = Grade A answer

Grade booster ····⟩ move A to A*
This is a good, solid answer from Heather, typical of the response expected from an A candidate. An A* candidate is also expected to give fully correct answers to some of the more difficult parts of the question. These usually occur towards the end.

Particles

1 In a diesel engine, air is drawn into a cylinder and then squashed rapidly. This causes the temperature to rise.

movement of piston

a) Explain why the pressure of the trapped air increases.

....

....

....

.... ④

b) Use the data in the table to calculate the pressure of the air that has been compressed by the moving piston.

	Air before compression	Air after compression
Pressure in kPa	100	
Volume in cm³	400	25
Temperature in K	290	700

....

....

.... ③

TOTAL 7

2 The diagram shows some of the possible results of firing alpha particles at thin gold foil.

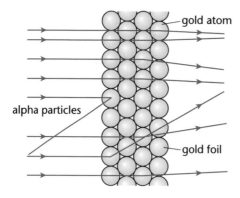

gold atom

alpha particles

gold foil

a) Describe three possible outcomes.

....

....

.... ④

b) Which piece of evidence:

  i) shows that the atom is mainly empty space

  .... ①

  ii) shows that there are intense regions of charge

  .... ①

  iii) shows that the nucleus has the same sign of charge as an alpha particle?

  .... ①

c) How did the results of alpha particle scattering change the accepted model of the atom?

  ....

  ....

  ....

  .... ④

TOTAL 11

3 The diagram shows an electron gun.

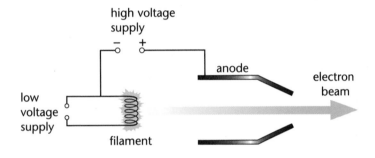

a) i) Explain how electrons are released from the filament.

  ....

  ....

  .... ③

  ii) What is the name of this effect?

  .... ①

b) What is the purpose of the anode?

  .... ①

c) The anode voltage is 450 V.
   The charge on an electron, $e = 1.60 \times 10^{-19}$ C.
   Calculate the kinetic energy of an electron as it leaves the electron gun.

  ....

  .... ②

d) State **two** devices that use electron beams.

  ....

  .... ②

TOTAL 9

4 The graph shows the relationship between the number of protons and the number of neutrons in stable nuclei.

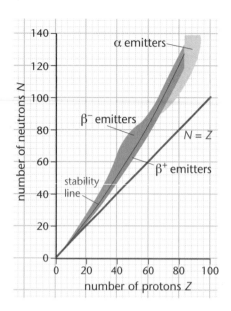

a) Explain what type of radiation is emitted when the isotope zirconium-86, $^{86}_{40}$Zr, decays.

....

....

.... ②

b) Complete the equation for the decay of zirconium-86.

$$^{86}_{40}\text{Zr} \longrightarrow \text{Y} +$$ ②

c) Use the graph to explain how this decay enables the nucleus to become more stable.

....

....

.... ②

d) What change takes place in the nucleus when zirconium-86 decays?

....

.... ②

TOTAL 8

5 The diagram shows a storage heater. The heating elements are turned on at night when electricity is cheap and heat is obtained from the concrete blocks during the day.

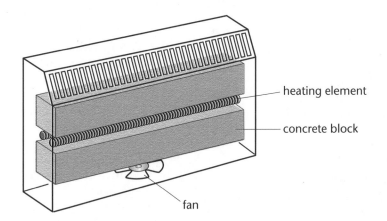

heating element

concrete block

fan

The heater contains 85 kg of concrete that has a specific heat capacity of 1200 J/kg °C.

a) How much energy is stored in the concrete when it is heated from 20°C to 200°C?

....

....

.... ③

b) Explain how this energy is stored by the particles that make up the concrete.

....

....

.... ③

TOTAL 6

1 a) The same number of particles is in a smaller volume ❶
Therefore collisions with the walls are more frequent ❶
The air is hotter so the particles are moving faster on average ❶
Therefore a bigger force is exerted at each collision ❶

Examiner's tip

It is important to stress that the increase in pressure is due to more frequent collisions between the gas particles and the container walls. A common error at GCSE is to explain that gas pressure is a result of collisions between particles.

b) $P_2 = \dfrac{P_1 V_1 T_2}{T_1 V_2}$ ❶

$= \dfrac{100 \text{ kPa} \times 400 \text{ cm}^3 \times 700 \text{ K}}{290 \text{ K} \times 25 \text{ cm}^3}$ ❶

$= 3860 \text{ kPa}$ ❶

Examiner's tip

The first mark here is for correct transposition of the gas equation. There is no mark for recall, as the relationship is given on the question paper or in a data booklet.

2 a) Some alpha particles pass straight through ❶
Some are deflected by varying amounts ❶
A tiny number are back-scattered ❶
Quality of written communication mark – all the information given is relevant to the question ❶

Examiner's tip

Take care to use the correct scientific terms. In answering this question some candidates lose marks because they describe the alpha particles as being 'reflected' rather than 'deflected'.

b) i) Most of the alpha particles are undeviated ❶
ii) The large deflection of a small number of the alpha particles ❶
iii) Some alpha particles are back-scattered ❶

Examiner's tip

Back-scattering is the only conclusive evidence that the nucleus has the same sign charge as an alpha particle.

c) The old model described the atom as being a uniform positively charged mass ❶
With negative charges evenly distributed ❶
The current model is a tiny nucleus that is positively charged and contains most of the mass ❶
The nucleus is orbited by negatively-charged electrons ❶

3 a) i) The filament is heated by the current in it ❶
The average energy of the particles increases ❶
Some electrons gain enough energy to leave the filament ❶

Examiner's tip

Once the electrons have left the filament, they have little energy. A common misunderstanding is that energy from the hot filament gives the electrons kinetic energy. This energy comes from the anode.

ii) Thermionic emission ❶
b) To accelerate and focus the beam ❶
c) Gain in kinetic energy
$= e \times V = 1.60 \times 10^{-19} \text{ C} \times 450 \text{ V}$ ❶
$= 7.2 \times 10^{-17} \text{ J}$ ❶

Examiner's tip

This relationship is provided on the question paper. Check your specification so that you know which relationships are given and which you need to learn. This varies between examination groups.

d) TV picture tubes
Computer monitors
Oscilloscopes
X-ray machines   any two, 1 mark each ❷

4 a) Zirconium-86 lies below the stability line ❶
So it emits β^+ radiation ❶
b) $^{86}_{40}\text{Zr} \rightarrow \,^{86}_{39}\text{Y} + \,^{0}_{1}\text{e}$

1 mark for each correct symbol on the right-hand side of the equation ❷

Examiner's tip

When writing a balanced nuclear equation, remember that the mass numbers must add up to the same on each side of the equation. This is also the case for the atomic numbers.

c) The new nucleus is to the left and higher up on the graph than the old one ❶
This brings it closer to the stability line ❶

d) A proton changes to a neutron by emitting a β⁺ particle ❶
This results in one less proton and one more neutron ❶

Examiner's tip

When a proton changes to a neutron an up quark becomes a down quark.

❺ a) Energy = mass × specific heat capacity × temperature change ❶
= 85 kg × 1200 J/kg °C × 180°C ❶
= 18 360 000 J ❶

Examiner's tip

It does not matter whether you work with temperatures in °C or in kelvin. A temperature change has the same value on both scales.

b) The energy of the concrete is the internal energy of its particles ❶
As the temperature increases, this internal energy increases ❶
On average, the particles have increased potential and kinetic energy ❶

Examiner's tip

The potential energy referred to here is not gravitational potential energy. It is the energy that is stored as two neighbouring particles move together.

Centre number	
Candidate number	
Surname and initials	

 **Examing Group**

General Certificate of Secondary Education

Physics
Paper 1

Higher tier

Time: One and a half hours

For Examiner's use only	
1	
2	
3	
4	
5	
6	
7	
Total	

Instructions to candidates

Answer all questions in the spaces provided.

The number of marks for each question is given in brackets at the end of each part.

These marks are a guide for the detail required in each answer.

Use a sharp pencil when drawing a graph or diagram.

In certain questions extra marks are available for the quality of your written answer.

Allocate a few minutes towards the end of the examination to check your answers.

Information for candidates

The number of marks available is given in brackets [2] at the end of each question or part question.

The marks allocated and the spaces provided for your answers are a good indication of the length of answer required.

EDUCATIONAL

1 An electric bar heater has two heating elements. Each has its own switch.

(a) The elements are connected in parallel. State **two** advantages of connecting the elements in parallel rather than in series.

....

.... **[2]**

(b) Explain whether each switch should be in series or in parallel with the element that it controls.

.... **[1]**

(c) One element is switched on. The voltage across the element is 240 V and the current passing in it is 4.0 A.

(i) Calculate the resistance of the element at its operating temperature.

....

....

.... **[3]**

(ii) Calculate the power of the element.

....

....

.... **[3]**

(iii) Explain why the element glows red when it is operating.

....

.... **[2]**

(d) The cost of energy from the electricity supply is 7p for each kWh.
The heater is used with both elements switched on for a time of 3.5 hours each day. Calculate the weekly cost of using the heater. Use the relationship:

cost = power in kW × time in hours × cost of 1 kW h

....

.... **[2]**

(Total 13 marks)

[turn over

Letts

2 **(a)** The diagram represents a wave travelling along a rope.

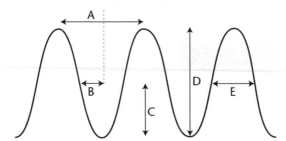

(i) Which arrow could represent the vibrations of the rope? Explain how you can tell.

....

.... **[2]**

(ii) Which arrow shows a distance equal to half a wavelength of the wave?

.... **[1]**

(iii) Which arrow shows a distance equal to the amplitude of the wave?

.... **[1]**

(iv) Explain why the rope cannot be used to model what happens when a sound wave is transmitted through the air.

....

.... **[2]**

(v) A wave of frequency 2.4 Hz passes along the rope. The wavelength of the wave is 0.35 m. Calculate the speed of the wave along the rope.

....

....

.... **[3]**

(b) A loudspeaker system used in a hi-fi set consists of two loudspeakers; one has a diameter of 6 cm and one a diameter of 30 cm.

The loudspeaker reproduces sounds that range in frequency between 20 Hz and 15 000 Hz.

(i) The speed of sound in air is 330 m/s.
Calculate the wavelength in air of a sound that has a frequency of 15 000 Hz.

....

.... **[3]**

(ii) Explain why a sound of this wavelength should be reproduced using the smaller loudspeaker.

....

.... **[2]**

(Total 14 marks)

[turn over

Letts

3 The diagram shows a satellite in orbit around the Earth.

(a) Add an arrow to the diagram to show the force on the satellite. **[1]**

(b) Which word from the list describes the type of force acting on the satellite?
Underline your choice. **[1]**

       electric   **gravitational**   **magnetic**   **nuclear**

(c) Give **two** uses of artificial satellites.

....

.... **[2]**

(d) The time it takes a satellite to complete one orbit of the Earth depends on its
height above the Earth's surface.
Some orbit times are given in the table.

Height above Earth's surface in millions of metres	Orbit time in hours
0	1.5
12	7.0
26	16.0
40	27.5
50	37.0

(i) Use the grid to draw a graph of height above the Earth's surface against
orbit time. **[3]**

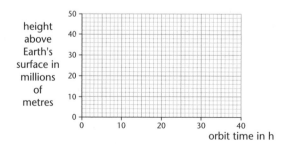

A satellite orbits the Earth at a height of 36 million metres.

(ii) Use the graph to find the orbit time of the satellite.

.... [1]

(iii) Explain why this satellite stays above the same point on the Earth's surface.

....

.... [2]

(iv) Suggest a use for this satellite.

.... [1]

(Total 11 marks)

[turn over

4 Technetium-99 is a radioactive isotope used in medicine. It decays by emitting gamma radiation.

Leave blank

(a) What is gamma radiation?

....

.... **[2]**

(b) The activity of a sample of technetium-99 is measured at intervals of two hours, starting when the sample has been prepared. The graph shows the results.

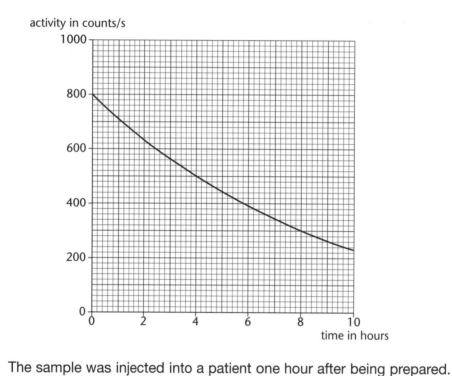

    (i) The sample was injected into a patient one hour after being prepared. What was the activity of the sample when it was injected into the patient?

.... **[1]**

   (ii) Use the graph to determine the half-life of technetium-99. Show how you obtain your answer.

....

.... **[2]**

  (iii) Estimate the activity of the sample one day after it is injected into the patient.

....

.... **[3]**

(c) Technetium-99 can be used to monitor the blood flow in organs such as the liver.

After the patient has been injected, a camera is used to photograph the patient from the outside.

Explain why a gamma emitter is used rather than a substance that emits alpha or beta particles.

....

....

.... **[3]**

(Total 11 marks)

[turn over

5 Two coils of wire are wound on an iron core.
One coil is connected to a battery and a switch.
The other coil is connected to a sensitive ammeter that can detect current in either direction.

(a) What happens to the ammeter pointer when:

    (i)   the switch is closed

    .... **[1]**

    (ii) the switch remains closed

    .... **[1]**

    (iii) the switch is opened?

    ....

    .... **[2]**

(b) The left-hand coil is now connected to an a.c. supply and the right-hand coil is connected to a lamp.

Explain why the lamp lights.

    ....

    ....

    .... **[3]**

(c) In an arrangement similar to that in **(b)**, a 3 V a.c. supply is connected to a 60-turn coil of wire.
The second coil is connected to a 12 V lamp, which lights at its normal brightness.
How many turns of wire are on the second coil?

....

....

.... **[3]**

(d) What is the name of the device described in **(c)**?

.... **[1]**

(Total 11 marks)

[turn over

6 A model rocket is fired vertically into the air.
The diagram shows the rocket's flight.

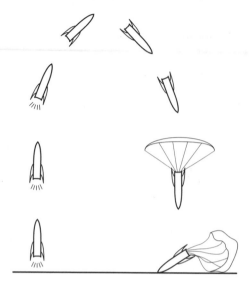

The graph shows how the velocity of the rocket changes during part of its flight.

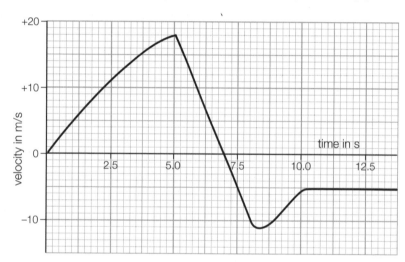

(a) (i) How does the acceleration of the rocket change between the times of 3 s
and 6 s?

....

.... **[2]**

(ii) Suggest what happened to cause this.

.... **[1]**

(b) During its ascent, the rocket travelled further while it was accelerating than while it was decelerating.
How can you tell this from the graph?

....

.... **[2]**

(c) At what time did the rocket begin to fall?
Explain how you can tell.

....

.... **[2]**

(d) After 8.5 s the parachute opened.

  (i) Describe the change in the speed of the rocket during the next 2 s.

  ....

  .... **[2]**

  (ii) Explain why the speed changed in this way.

  ....

  ....

  .... **[3]**

  (iii) What is the value of the rocket's terminal velocity?

  .... **[1]**

(Total 13 marks)

[turn over

7 A car accelerates from 20 m/s to 35 m/s in 6.0 s.

(a) Calculate the acceleration of the car.

....

....

.... **[3]**

(b) The car and driver have a total mass of 950 kg.
Calculate the force needed to accelerate the car.

....

....

.... **[3]**

(c) Calculate the kinetic energy when the car is travelling at 35 m/s.

....

....

.... **[3]**

(d) The car brakes. The brakes remove energy at the rate of 75 000 J/s.
Calculate the time it takes for the brakes to stop the car.

....

.... **[2]**

(e) How far does the car travel during braking?

....

....

.... **[3]**

(f) When the car is fully laden its mass is 1250 kg.
Explain how this affects the acceleration and the braking of the car.

....

....

.... **[3]**

(Total 17 marks)

Letts **Examining Group**

General Certificate of Secondary Education

Physics
Paper 2

Higher tier

Time: One hour

Instructions to candidates

Write your name, centre number and candidate number in the boxes at the top of this page.

Answer ALL questions in the spaces provided on the question paper.

Show all stages in any calculations and state the units. You may use a calculator.

Include diagrams in your answers where this may be helpful.

Information for candidates

The number of marks available is given in brackets **[2]** at the end of each question or part question.

The marks allocated and the spaces provided for your answers are a good indication of the length of answer required.

Letts
EDUCATIONAL

1 The diagram shows a potential divider circuit.

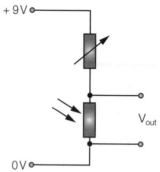

+ 9V

V_{out}

0V

(a) Describe the function of this circuit.

....

.... **[2]**

(b) Calculate the value of the output voltage, V_{out}, when the resistance of the variable resistor is 975 Ω and that of the LDR is 150 Ω.

....

.... **[3]**

(c) Explain how the output voltage changes as the intensity of the light increases.

....

.... **[2]**

(d) The output voltage causes a switching circuit to operate when the voltage across it is 0.60 V.

(i) The value of the variable resistor is maintained at 975 Ω.
What is the resistance of the LDR when the switch operates?

....

.... **[2]**

(ii) The light intensity is maintained at the level at which the resistance of the LDR is 150 Ω.
The value of the variable resistor is adjusted.
What is the resistance of the variable resistor when the switch operates?

....

.... **[2]**

(Total 11 marks)

2 **(a)** The refractive index of glass is 1.50.
The speed of light in air is 3.00×10^8 m/s.
Calculate the speed of light in glass.

....

....

.... **[3]**

(b) **(i)** Explain what is meant by the dispersion of light.

.... **[1]**

(ii) Explain why light is sometimes dispersed as it passes from air into glass.

....

.... **[2]**

(c) An object is placed 8.0 cm from a concave lens.
The height of the object is 2.5 cm.
The focal length of the lens is 10.0 cm.

(i) Complete the ray diagram to find the position and properties of the
image. **[2]**

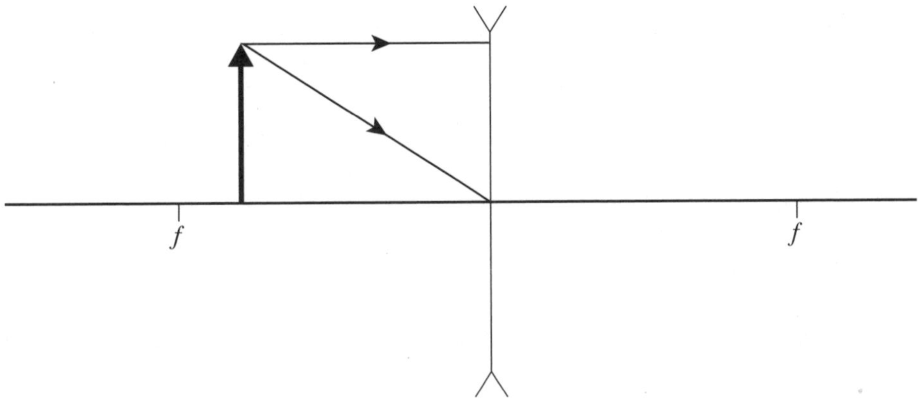

(ii) Cross out the words that do not apply to the image.
The image is real/virtual and upright/inverted. **[1]**

(iii) Complete the sentences by taking measurements from your completed
diagram.

The height of the image is .... cm and

it is .... cm from the lens. **[1]**

(Total 10 marks)

[turn over

Letts

3 The diagram shows three ways in which radio waves can travel.

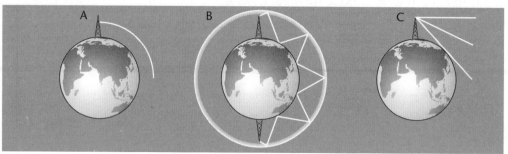

(a) What type of wave is shown in diagram:

A _____

B _____

C? _____ **[2]**

(b) Explain which type of wave is used for satellite transmissions.

....

.... **[2]**

(c) Modern satellite communications use digital signals.
Like analogue signals, they are affected by noise, but they can be regenerated.
Draw a diagram to show how a digital signal is affected by noise and how the
signal is changed when it is regenerated.

[2]

(d) Weather satellites often occupy low polar orbits, but communications satellites
are often in geostationary orbits.

(i) Describe the differences between a low polar orbit and a geostationary orbit.

....

....

.... **[3]**

(ii) Why is it an advantage to have a geostationary orbit for a communications
satellite?

.... **[1]**

(Total 10 marks)

4 A moving vehicle has both momentum and kinetic energy.

(a) (i) Which of these quantities are conserved in an elastic collision?

.... **[1]**

(ii) Which of these quantities are conserved in an inelastic collision?

.... **[1]**

(b) The diagram shows what happens when two 'vehicles' collide on an air track.

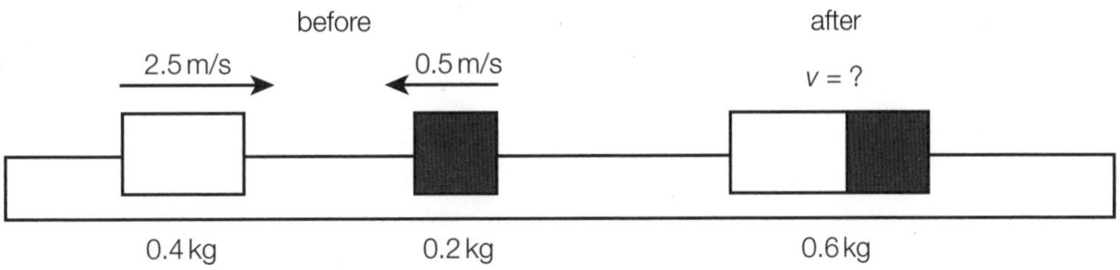

(i) Calculate the total momentum of the vehicles before the collision.

....

....

....

.... **[4]**

(ii) Calculate the velocity of the vehicles after the collision.

....

....

.... **[2]**

(c) Explain why the results of air track experiments give results that are in closer agreement with the principle of conservation of momentum than experiments that are carried out on steel track with no air cushion.

....

.... **[2]**

(Total 10 marks)

[turn over

Letts

5 The Moon is the Earth's natural satellite.
Here is some data about the Earth's moon:

radius of orbit = 3.8×10^5 m
orbital speed = 1.0 m/s
mass = 7.4×10^{22} kg

(a) Calculate the time that it takes the Moon to complete one orbit of the Earth.

....

.... **[2]**

(b) (i) Calculate the centripetal acceleration of the Moon.

....

.... **[2]**

(ii) Describe the force that causes this acceleration.

....

.... **[2]**

(iii) Calculate the value of the centripetal force acting on the Moon.

....

.... **[2]**

(iv) State the size and direction of the force that the Moon exerts on the Earth.

....

.... **[2]**

(Total 10 marks)

6 The diagram shows the trace on the screen of an oscilloscope.

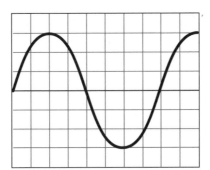

(a) (i) How can you tell from the trace that the time-base is switched on?

.... **[1]**

(ii) The time-base is now switched off.
Describe the appearance of the screen.

....

.... **[2]**

(b) The y-sensitivity is set to 2.5 V/cm.
The time-base control is set to 0.005 s/cm.

(i) Calculate the maximum value of the input voltage.

.... **[1]**

(ii) Calculate the minimum value of the input voltage.

.... **[1]**

(iii) Calculate the frequency of the input voltage.

....

....

.... **[2]**

[turn over

Letts

(c) An oscilloscope uses the electric field between two charged plates to deflect the electron beam.
The diagram shows the electron beam as it is about to enter the region between the charged plates.

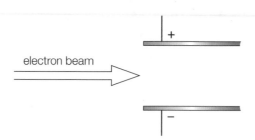

Draw a line to show the path of the electron beam as it passes between and out of the electric field. **[2]**

(Total 9 marks)

Mock examination Paper 1 answers

1 (a) The elements can be switched independently [1]

Both have the full mains voltage [1]

Examiner's tip

When components are connected in series, the supply voltage is shared between them.

(b) In series so it breaks the circuit. In parallel it would short-circuit. [1]

(c) (i) Resistance = voltage ÷ current [1]

= 240 V ÷ 4.0 A [1]

= 60 Ω [1]

(ii) Power = voltage × current [1]

= 240 V × 4.0 A [1]

= 960 W [1]

(iii) Electric currents cause heating [1]

If a wire becomes hot enough, it also gives out light [1]

Examiner's tip

For a given current, the rate of heating is proportional to the resistance. This is why connecting wires should always have a low resistance.

(d) Cost = 2 × 0.96 kW × 24.5 h × 7p/kW h [1]

= 329p or £3.29 [1]

Examiner's tip

If you choose to change the cost from p to £, make sure that you do it correctly. A common error in this question is to give the final answer as £329 or £329.28 – the figures that appear on the calculator display.

2 (a) (i) D [1]

The wave shown is transverse [1]

(ii) E [1]

Examiner's tip

B shows a distance equal to one quarter of a wavelength and A shows one wavelength. One wavelength is one complete cycle of a wave, a crest and a trough for a transverse wave or a compression and a rarefaction for a longitudinal wave.

(iii) C [1]

Examiner's tip

A common error in this question is to answer D. The amplitude is the maximum displacement from the mean or rest position, not the difference between the extremes of displacement.

(iv) Sound is a longitudinal wave [1]

a rope cannot easily demonstrate longitudinal waves [1]

(v) Speed = frequency × wavelength [1]

= 2.4 Hz × 0.35 m [1]

= 0.84 m/s [1]

(b) (i) Wavelength = speed ÷ frequency [1]

= 330 m/s ÷ 15 000 Hz [1]

= 0.022 m [1]

Examiner's tip

The first mark here is for a correct transposition of a relationship that is required recall (speed = frequency × wavelength).

(ii) The wavelength is close to the size of the small speaker/the large speaker is many times the wavelength [1]

The small speaker gives more spreading out of the sound [1]

3 (a) The arrow should point towards the centre of the Earth [1]

(b) Gravitational [1]

Examiner's tip

Planets, moons and satellites are kept in orbit around a larger body because of the attractive gravitational forces between them. These forces act from centre to centre of the two objects. A common error is to draw the force arrow in the direction of motion.

(c) Surveillance, Weather, Navigation, Studying space   any two, 1 mark each [2]

(d) (i)

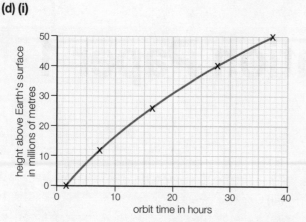

Correct plot (allow one error) **[2]**

award one mark if there are two or three errors

Curve drawn **[1]**

(ii) 24 hours **[1]**

(iii) This is the same as the time it takes the Earth to rotate once on its axis **[2]**

award 1 mark for identifying this time with the length of a day

Examiner's tip

A common error here is to state that the satellite rotates at the same speed as the Earth. This is not the case. The satellite actually moves faster than any point on the Earth's surface as it is further from the axis of rotation.

(iv) For transmitting television/telephone/other communications **[1]**

4 (a) Short wavelength or high frequency **[1]**

Electromagnetic radiation **[1]**

(b) (i) 705 counts/s **[1]**

(ii) One reading in range 5.7 to 5.8 hours shown taken from the graph **[1]**

A second reading taken and the average calculated **[1]**

Examiner's tip

When taking readings from a graph, always show clearly on the graph how you have done this. In this case a horizontal line should be drawn across to the curve from half the initial value of the activity. Then a vertical line should be drawn down to the time axis.

(iii) One day is approximately four half-lives **[1]**

The activity will be $\frac{1}{2} \times \frac{1}{2} \times \frac{1}{2} \times \frac{1}{2}$ of 700 counts/s **[1]**

= 43 counts/s **[1]**

(c) Gamma radiation can penetrate the flesh and be detected by the photographic film in the camera **[1]**

Alpha and beta radiation would be absorbed in the body **[1]**

And could damage cells **[1]**

Examiner's tip

Both alpha and beta radiation cause more ionisation than gamma radiation does. They can therefore cause a lot more damage than gamma radiation to body cells and tissue.

5 (a) (i) It moves/shows a reading and then returns to zero **[1]**

(ii) The pointer stays at zero **[1]**

(iii) It moves/shows a reading and then returns to zero **[1]**

The movement/reading is in the opposite direction to that in **(a) (i)** **[1]**

Examiner's tip

A voltage or current is only induced in the right-hand coil when the magnetic field is changing. This happens when a direct current is switched on or off. Switching the current off causes an induced voltage in the opposite direction to when it is switched on.

(b) The current is continually changing **[1]**

So the magnetic field is continually changing **[1]**

A voltage is induced whenever the magnetic field changes **[1]**

Examiner's tip

Always emphasise the changing magnetic field when explaining the effects of electromagnetic induction.

(c) $\dfrac{V_p}{V_s} = \dfrac{N_p}{N_s}$ **[1]**

$\dfrac{3V}{12V} = \dfrac{60}{N_s}$ **[1]**

$N_s = 240$ **[1]**

Examiner's tip

This type of question is easier to answer by using ratios than by using the transformer equation. If the secondary voltage is four times that of the primary, the secondary coil must have four times as many turns as the primary coil.

(d) A transformer **[1]**

6 (a) (i) The acceleration decreases [1]

And then changes direction [1]

Examiner's tip

The gradient of a velocity–time graph represents acceleration in both size and direction.

(ii) The rocket ran out of fuel [1]

Examiner's tip

Take care when interpreting velocity–time graphs. After 5.0 s the rocket is still travelling upwards, although its acceleration is now in the downwards direction. This is because it is slowing down.

(b) The area between the curve and the time axis represents the distance travelled [1]

This area is greater for the first 5 s than for the next 2 s [1]

(c) 7.0 s [1]

The velocity of the rocket reversed from a positive velocity to a negative one [1]

(d) (i) The speed decreases [1]

And then maintains a constant value [1]

(ii) When the parachute opened the upward force of air resistance increased [1]

The upward force was greater than the downward force, so the speed decreased [1]

It maintained a constant value when the forces were balanced [1]

Examiner's tip

Unbalanced forces cause a change in speed or direction or both. There is no change in speed or direction when the forces on an object are balanced.

(iii) −5 m/s. [1]

7 (a) Acceleration

= increase in velocity ÷ time taken [1]

= 15 m/s ÷ 6 s [1]

= 2.5 m/s^2 [1]

Examiner's tip

Take care with the unit when working out accelerations. Most candidates at GCSE give the unit as m/s, losing one mark out of three.

(b) Force = mass × acceleration [1]

= 950 kg × 2.5 m/s^2 [1]

= 2375 N [1]

(c) Kinetic energy = $\frac{1}{2}$ × mass × (speed)^2 [1]

= $\frac{1}{2}$ × 950 kg × (35 m/s)^2 [1]

= 581 875 J [1]

Examiner's tip

Common errors when calculating kinetic energy are forgetting to square the speed and forgetting the half.

(d) Time = 581 875 J ÷ 75 000 s [1]

= 7.76 (7.8) s [1]

(e) Distance travelled = average speed × time [1]

= 17.5 m/s × 7.76 s [1]

= 136 m [1]

Examiner's tip

If your answer to (e) is 272 m, you have not taken account of the fact that in braking from 35 m/s to rest (0 m/s), the average speed of the car is half of the initial speed.

(f) The acceleration is reduced [1]

The deceleration is reduced [1]

The braking or stopping distance is increased [1]

Mock examination Paper 2 answers

1 (a) It provides an output voltage that varies with light intensity [1]

The value of the voltage at any particular intensity can be adjusted with the variable resistor [1]

(b) $V_{out} = V_{in} \times \dfrac{R_2}{R_1 + R_2}$ [1]

$= 9\,V \times 150\,\Omega \div (975\,\Omega + 150\,\Omega)$ [1]

$= 1.2\,V$ [1]

Examiner's tip

This is a cumbersome relationship to use. It is much easier to work out the voltages using ratios. For resistors in series, the ratio of the voltages is the same as the ratio of the resistances.

(c) The resistance of the LDR decreases [1]

So the voltage across it decreases [1]

Examiner's tip

This circuit would be useful for switching on an exterior light at night, as the voltage across the LDR rises when the light intensity falls.

(d) (i) $0.60\,V = 9\,V \times \dfrac{R_2}{975\,\Omega + R_2}$ [1]

$R_2 = 69.6\,\Omega$ [1]

Examiner's tip

Using ratios, you can see that the voltage across the LDR is 1/14 of that across the variable resistor, so its resistance must be 1/14 of the variable resistor's resistance.

(ii) $0.60\,V = 9\,V \times \dfrac{150\,\Omega}{R_1 + 150\,\Omega}$ [1]

$= 2100\,\Omega.$ [1]

2 (a) Speed of light in glass = speed of light in air ÷ refractive index [1]

$= 3.00 \times 10^8\,m/s \div 1.5$ [1]

$= 2.00 \times 10^8\,m/s.$ [1]

Examiner's tip

There is very little difference between the speeds of light in air and in a vacuum; the same value is usually used for both.

(b) (i) Light being split up into different colours [1]

(ii) Different colours/frequencies of light have the same speed in air but different speeds in glass [1]

So when the light changes direction, the different colours change direction by different amounts [1]

(c) (i) The light parallel to the axis should be continued as if it had come from the principal focus on the left of the lens. [1]

This ray is traced back to the tip of the image, where it crosses the ray directed towards the centre of the lens. [1]

Examiner's tip

When drawing ray diagrams for concave lenses, remember that light parallel to the axis diverges away from the principal focus.

(ii) The image is virtual and upright [1]

(iii) 1.4 and 4.5 [1]

3 (a) A ground wave

B sky wave

C space wave

three correct [2]

allow 1 mark for 2 correct

(b) C/space wave [1]

As it is the only one that can pass through the atmosphere [1]

(c) Here is the completed diagram

a digital signal   noise   regeneration

Marks are awarded for:

Noise affecting the amplitude of the digital signal only [1]

Regeneration showing the digital signal restored [1]

Examiner's tip

Noise does not affect the frequency of a signal, only its amplitude. This is why frequency modulation is used to carry high-quality signals.

(d) (i) A polar orbit is over the poles, but a geostationary orbit is over the equator [1]

A low polar orbit passes over a different section of the Earth on successive orbits, but a geostationary orbit is fixed relative to the Earth [1]

A low polar orbit has a much shorter time period than that of a geostationary orbit [1]

(ii) Because it remains at a fixed point above the Earth's surface, it can be used 24 hours a day with fixed aerials [1]

Examiner's tip

Note that question (d)(i) is about orbits, but (d)(ii) is about a satellite in orbit. Take care to make sure which of these the question is asking about.

4 (a) (i) Momentum and kinetic energy [1]

(ii) Momentum only [1]

Examiner's tip

Momentum is always conserved and total energy is always conserved. In an inelastic collision there is some energy transfer from kinetic to other forms – mainly heat.

(b) (i) Momentum = mass × velocity [1]
= 0.4 kg × 2.5 m/s [1] − 0.2 kg × 0.5 m/s [1]
= 0.9 kg m/s [1]

Examiner's tip

Remember that momentum is a vector quantity, so the direction needs to be taken into account. In this case the direction from left to right has been taken as positive. An alternative unit for momentum is the Ns.

(ii) Velocity = momentum ÷ mass
= 0.9 kg m/s ÷ 0.6 kg [1]
= 1.5 m/s from left to right. [1]

Examiner's tip

It is important to state the direction for the second mark.

(c) The principle of conservation of momentum applies to colliding objects where there are no other forces acting on them [1]

Friction with the steel surface is a significant 'other force' but air resistance has a much smaller effect. [1]

5 (a) Time = $2\pi r \div v$
= $2 \times \pi \times 3.8 \times 10^5$ m ÷ 1.0 m/s [1]
= 2.39×10^6 s. [1]

(b)(i) Acceleration = $v^2 \div r$
= $(1.0 \text{ m/s})^2 \div 3.8 \times 10^5$ m [1]
= 2.63×10^{-6} m/s^2 [1]

Examiner's tip

This is the value of the Earth's gravitational field strength at the Moon's orbit.

(ii) The pull of the Earth [1]
On the Moon [1]

Examiner's tip

No marks are awarded for 'gravity'. This is an ambiguous term that is best avoided, as the same word is used to mean force, acceleration and gravitational field strength. This leads to confusion.

(iii) Force = mass × acceleration
= 7.4×10^{22} kg × 2.63×10^{-6} m/s^2 [1]
= 1.95×10^{17} N. [1]

(iv) 1.95×10^{17} N [1]
Towards the centre of the Moon [1]

Examiner's tip

When objects exert forces on each other, the forces are equal in size and opposite in direction.

6 (a) (i) The 'dot' appears as a horizontal trace [1]

(ii) A vertical line [1]
6 cm long [1]

Examiner's tip

With the time-base switched off, the dot is only moving vertically.

(b) (i) +7.5 V [1]

(ii) −7.5 V [1]

Examiner's tip

The trace shows an alternating voltage varying between +7.5 V at the peak to −7.5 V at the trough.

(iii) Time period = 8 cm × 0.005 s/cm

                  = 0.040 s   **[1]**

    Frequency = 1 ÷ time period

                = 1 ÷ 0.040 s = 25 Hz.   **[1]**

Examiner's tip

A common error is to measure the time for half of the cycle rather than one complete cycle of the alternating voltage.

(c) The diagram should show:

    The electron beam curves upwards as it passes between the plates   **[1]**

    And continues in a straight line when it has left the area between the plates   **[1]**

Examiner's tip

When the electrons have left the region of the electric field there is no longer a force on them, so they continue straight. Because of their high speed, the effect of the Earth's gravitational field is not noticeable.

Index

Index